Principles and Methods of Reclamation Science

Principles & Methods of Reclamation Science

with

Case Studies from the Arid Southwest

EDITED BY CHARLES C. REITH
AND LOREN D. POTTER

University of New Mexico Press
Albuquerque

Library of Congress Cataloging-in-Publication Data
Main entry under title:

Principles and methods of reclamation science.

   Includes bibliographies and index.
   Contents: Introduction and overview of methods
in reclamation science / Charles C. Reith—
Applications of geomorphology to reclamation /
Stephen G. Wells and Loren D. Potter—Pre-mining
assessments of reclamation potential / Loren D.
Potter—[etc.]
   1. Reclamation of land.   2. Strip mining—
Environmental aspects.   3. Revegetation.
4. Reclamation of land—San Juan Basin (N.M. and
Colo.)   5. Strip mining—Environmental aspects—
San Juan Basin (N.M. and Colo.)   6.
Revegetation—San Juan Basin (N.M. and Colo.)
I. Reith, Charles C. II. Potter, Loren D.
S621.5.S8P75 1986   631.6′4′0978982   85–24645
ISBN 0-8263-0830-9
ISBN 0-8263-0831-7 (pbk.)

Brief Table of Contents

Complete Table of Contents

Tables

Figures

Contents

1

Introduction
and Overview of Methods
in Reclamation Science

Charles C. Reith

Objectives of the Book

Reclamation science is a dispersed science, with elements in geology, ecology, agronomy, and other disciplines. At present, reclamation research is dispersed among these disciplines, and interdisciplinary communication is poor. In designing, conducting, and interpreting research, reclamation scientists tend to restrict their thinking to the facts and theories within their disciplines. In developing reclamation policies and programs, managers in government and industry tend to subscribe only to research from the sciences in which they were trained, disregarding the wealth of potentially valuable information from other sciences.

The principal goal and unique contribution of this book is to remedy the dispersed condition of reclamation science by providing the information which will enable reclamation scientists and managers in all disciplines to *synthesize* the science in their mind. This book strives to destroy the communication barrier—not by presenting the information in all the disciplines involved in our science, but by providing the intellectual tools that will equip scientists and managers from diverse backgrounds to assess and employ the information from all disciplines.

Many symposia and publications have recognized the multidisciplinary nature of reclamation, and have presented lectures, chapters, and contributions from all facets of the science. However, authors have not been encouraged to explain the rationale and jargon behind their specialty; thus, their presentations are often inaccessible to participants at all levels and disciplines in reclamation. The result is that scientists attend those sessions and read only those contributions which lie within their specialty. The worst result of the communication barrier is that some scientists will not only disregard but discredit the contributions of research which they do not understand.

Attacking the communication barrier in reclamation science

The chapters in this book are written by researchers and managers from all disciplines, but each author has been directed to open his science to all the participants in reclamation. The authors were encouraged to avoid the jargon and technical detail that has rendered their science inaccessible to others in reclamation. Another measure to achieve accessibility was to structure the chapters as consistently as possible, with each containing the following elements in sequence:

1. An identification and definition of the problem in reclamation which has motivated or necessitated the scientific approach in the chapter
2. An elaboration of the scientific approach or methodology, including its underlying premises, assumptions, advantages, and limitations
3. A case study of methodology in action, illustrating the details and logistics behind the method, as well as some representative findings
4. A brief review of related studies, illustrating the range of applications and findings of the methodology
5. A discussion which synthesizes the general findings and unanswered questions associated with the methodology, including a forecast of its future application to reclamation

These elements are not always discussed by using the exact terminology or sequence presented above, because different sciences and methodologies do not readily lend themselves to identical treatment. However, the authors and editors strove for the maximum possible consistency in the organization, scope, and detail in each chapter.

The result of our effort is a "three-way hybrid" between (1) a textbook describing the methods and principles of reclamation science, (2) a multidisciplinary compendium of reclamation research projects, and (3) a comprehensive review of the literature pertaining to land reclamation.

The San Juan Basin as a case study

The methods and principles presented herein apply to reclamation research worldwide. However, no compendium nor literature review of reclamation science could ever be exhaustive. Research projects are happening everywhere, each addressing the specific problems and constraints imposed by the local climate, geology, and land-use objectives. Indeed, another factor behind the communication barrier in reclamation science is the overwhelming proliferation of publications in journals and symposia sponsored by government, industry, and academe.

All of the case studies and most of the literature cited in this book pertain to the San Juan Basin, a discrete physiographic province in northwestern New Mexico and southwestern Colorado. We focus upon the Basin in order to illustrate how studies in all disciplines, regardless of their methods and objectives, must be tailored to the physical and biotic environment of the study area. Research must also address the principal land uses for the native landscape, which in the San Juan Basin is the provision of vegetation for livestock and wildlife, that is, the provision of rangeland.

On a more practical level, the emphasis upon the San Juan Basin means that background information for the case studies, including a detailed description of the Basin environment, can be given just once, here in the opening chapter. Otherwise, each case study would have to be preceded with a discussion of the background and environment for that study.

The emphasis on surface mining

The book also focuses upon a particular kind of reclamation, specifically the revegetation and stabilization of land which has been disturbed by surface mining. Although mine reclamation is probably the dominant theme of most current reclamation research, there are many other human impacts which require the restoration of disturbed lands. The broader field of land rehabilitation includes the reforestation of lands which have been burned or logged, the restoration of productivity to overgrazed rangelands, the reversal of desertification, and so forth.

There are also alternatives to revegetation as a rehabilitation option. Disturbed land may be landscaped and surfaced for specialized land uses such as airports, subdivisions, or industrial parks.

These various management challenges no doubt require different information, which must be provided by different kinds of research. But again, the principles and methods presented in the chapters and illustrated in the case studies are broadly applicable. Despite the diversity of approaches, all reclamation research addresses a few common elements: minimizing erosion, accelerating soil formation, and encouraging the germination, growth, and reproduction of vegetation. An understanding of these few elements, and of the ways to study them, will equip a researcher to conduct meaningful projects in any environment, on any aspect of reclamation.

The San Juan Basin

The San Juan Basin has a relatively long history of mining, reclamation, and research. The number and variety of reclamation research projects is due partly to the number of surface mines in the Basin, but more to the enormous amount of mining projected for the near and distant future.

The Basin has some uranium and other heavy metals, but the principal minable resource is subbituminous coal. Although this coal is high in ash, it is low in sulfur and fairly close to the surface, both of which make it attractive as a power source for electrical generating stations in and around the Basin. Grant (1981) predicts that approximately $4km^2$/year of land will be mined and reclaimed in the Basin's future.

Geography and environment

The San Juan Basin is a physiographic subdivision of the Colorado Plateau. The physiographic boundaries portrayed in Fig. 1.1 delimit a geographic area with a moderate degree of environmental variability. Figure 1.1 also locates the four surface coal mines and the single uranium mine which are mentioned in the case studies and which occupy nearly the full range of geologic and climatic conditions. This range, which depends mostly upon elevation, is illustrated by the data for vegetation, climate, and soils listed below the map.

The coal in the Basin is in two major geologic units—the Fruitland Formation and the Mesa Verde Sandstone—illustrated in Fig. 1.1. The coal was deposited in the Cretaceous period beneath brackish swamps surrounding an inland sea (Fassett and Hinds 1971). Many of the sedimentary rocks and derived soils in coal-bearing parts of the Basin are high in sodium and other soluble salts, inhibiting vegetative growth in the arid to semiarid climate.

The Basin has a "high desert" climate and vegetation (Bieri 1977). The average annual precipitation ranges from less than 15 cm, at the "Four Corners Area," to 40 cm or more at higher elevations near the Basin's edges. Most rain falls during torrential summer thunderstorms which exceed the infiltration rates of many soils, resulting in potentially erosive runoff. The poor infiltration of water into some soils and the erratic distribution of precipitation limit the availability of water to vegetation. Frequent sun and wind, plus hot afternoon temperatures, cause high potential evaporation, exceeding 125 cm over a growing season (Gould et al. 1975).

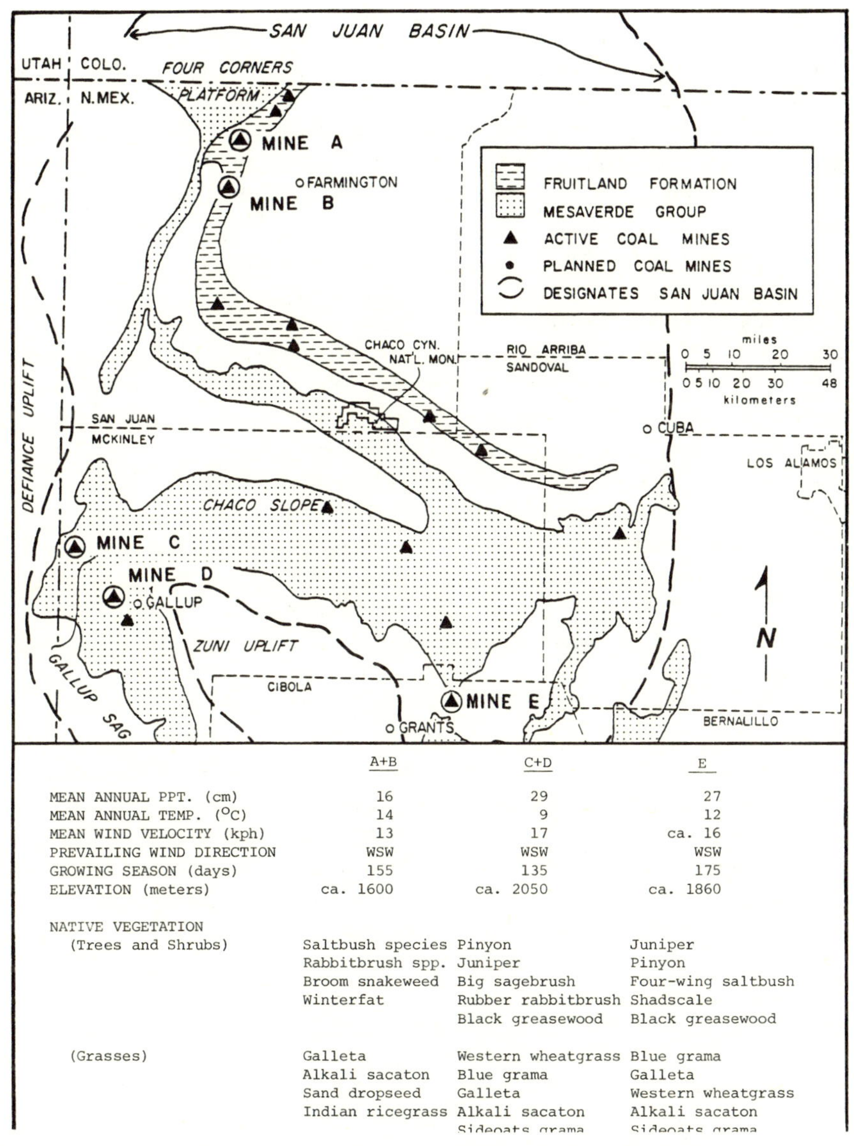

	A+B	C+D	E
MEAN ANNUAL PPT. (cm)	16	29	27
MEAN ANNUAL TEMP. ($^{\circ}$C)	14	9	12
MEAN WIND VELOCITY (kph)	13	17	ca. 16
PREVAILING WIND DIRECTION	WSW	WSW	WSW
GROWING SEASON (days)	155	135	175
ELEVATION (meters)	ca. 1600	ca. 2050	ca. 1860

NATIVE VEGETATION			
(Trees and Shrubs)	Saltbush species	Pinyon	Juniper
	Rabbitbrush spp.	Juniper	Pinyon
	Broom snakeweed	Big sagebrush	Four-wing saltbush
	Winterfat	Rubber rabbitbrush	Shadscale
		Black greasewood	Black greasewood
(Grasses)	Galleta	Western wheatgrass	Blue grama
	Alkali sacaton	Blue grama	Galleta
	Sand dropseed	Galleta	Western wheatgrass
	Indian ricegrass	Alkali sacaton	Alkali sacaton
	Sideoats grama	Sideoats grama	

Figure 1.1. Map of the San Juan Basin. (from Gutierrez 1979.)

6   Charles C. Reith

Soils and vegetation

The effect of the Basin's dry climate on vegetation is most severe where soils are rich in clay, for instance, soils derived from the shales of the Fruitland Formation. Water infiltrates more slowly into clay soils than sandy soils, thus more is lost to runoff. Furthermore, clay soils retain moisture on the surfaces of the small soil particles, causing plants to wilt even when water comprises as much as 15% of the soil dry weight (Brady 1974).

The adverse nature of clay soils in arid environments is exacerbated by salts in the soil, especially by sodium. Soluble salts (in *saline* soils) osmotically retain moisture in the soil, and sodium (in *sodic* soils) tends to destroy soil structure, limiting the infiltration of water and the aeration of roots. Clay-rich saline and sodic soils in the Basin are often barren of vegetation, except for a sparse covering of wild buckwheat *(Eriogonum* spp.) and saltbushes *(Atriplex* spp.) The lack of stabilizing vegetation and excessive runoff causes slopes with such soils to erode quickly into scenic badland formations (Fig. 1.2).

Clay soils in valleys and flatlands are dominated by big sagebrush *(Artemisia tridentata)*, rabbitbrush *(Chrysothamnus* spp.), black greasewood *(Sarcobatus vermiculatus)*, and perennial grasses such as western wheatgrass *(Agropyron smithii)*, galleta *(Hilaria jamesii)*, and alkali sacaton *(Sporobolus airoides)*.

Sandy soils support the highest coverages of vegetation in the Basin, occasionally exceeding 25%, comprised principally of four-wing saltbush *(Atriplex canescens)* and perennial range grasses such as Indian ricegrass *(Oryzopsis hymenoides)*, sand dropseed *(Sporobolus cryptandrus)*, and galleta. There are a few areas with sand dunes.

Mesa tops and rocky slopes in the Basin support a woodland of pinyon *(Pinus edulis)*, juniper *(Juniperus* spp.), and Gambel oak *(Quercus gambelii)*, with a mixture of grasses in the understory.

Land uses

Surface mining is only one land use in a complex array of alternatives for the San Juan Basin. The federal and state government faces an unprecedented management challenge because of the variety of land uses, the intensity of resource conflicts, and the complexity of land-ownership patterns, which involve several state and federal agencies, Indian tribes, and private interests. In fact, coal leasing in the San Juan Basin has been postponed several times because of the government's inability to resolve management conflicts.

Figure 1.2.   Badlands landscape in the San Juan Basin. These barren landscapes occur at outcrops of saline or sodic shales, where water infiltration is poor and erosion proceeds quickly.

The most prevalent land use is grazing for livestock, and the principal goal of reclamation is to reestablish range vegetation to a condition and value comparable to that before mining. If reclamation succeeds, surface mining only temporarily disrupts the value of the land for grazing. However, a number of land uses in the Basin are more directly in conflict with mining. These include the development of oil and gas, both of which occur below coal-bearing strata. Scientific resources such as paleontological and archaeological sites may be destroyed. The clear air which, along with the Basin's scenic landscapes and Indian ruins, attract tourists to the "Four Corners Area," may be marred by fugitive dust from mining and by pollution from power plants. Several particularly beautiful or unusual landscapes in the Basin are Wilderness Study Areas, and hence may become Wilderness Areas where mining is permanently precluded.

A final important factor is the presence of Indians of three tribes—the Navajo, Hopi, and Jicarilla Apache—on much of the coal-bearing land. Many of these longtime residents have life-styles requiring privacy and adherence to tradition. Widespread mining will have adverse cultural and physical impacts on the Indians and their land in the San Juan Basin.

Management decisions about mining must consider the economics of reclamation. In turn, the extent, feasibility, and cost of reclamation depends on the numerous land uses and resource conflicts mentioned above.

History of surface mining and regulation

Several surface mines have been in the Basin longer than the current mining and reclamation regulations. Reclamation on older surfaces at these mines was done on a voluntary and experimental basis, employing certain shortcuts and economies that are illegal today. These surfaces offer an opportunity to observe not only the durability of reclamation, but also the results of various shortcuts which may eventually be reintroduced to cut costs.

The most important legislation on the surface mining of coal in the United States was the Surface Mining Control and Reclamation Act of 1977 (SMCRA, called "smack-rah"). Details of SMCRA will be cited throughout this book; however, the thrust was to anticipate and mitigate (alleviate) the impacts of the surface mines on all aspects of the environment. Furthermore, the reclamation plan must adhere to specific criteria and must provide reasonable proof to regulators that revegetation and stabilization will be successful. Detailed data about the environment of the mine and the reclamation program are provided to regulators in a *mine plan*, which is reviewed and, if approved, forms the basis for a *mine permit*. Regulators periodically inspect the operation of the mine and the progress of reclamation to verify compliance with the terms of the permit.

With the passage of SMCRA in 1977, existing coal mines were required to prepare mine plans and to modify their reclamation programs to comply with regulations. The regulations were initially administered and enforced by the Office of Surface Mining (OSM), a branch of the Department of the Interior created by SMCRA. Responsibilities have been gradually and partially transferred to a state regulatory agency, specifically the Mining and Minerals Division of the New Mexico Energy and Minerals Department. The state agency has since enacted new regulations that recognize some unique problems of the arid Southwest, which will be addressed throughout this book.

Surface coal mines are planned throughout the coal-bearing regions of the San Juan Basin. Some have mine plans which have been approved or which are being prepared. However, much of the coal in the Basin is in leases which remain unsold or which contain legal barriers to mining. The sale of these leases and the resolution of legal problems is an arduous procedure that prevents mining from becoming too widespread in the Basin. The ultimate regulator of the pace at which coal is developed is the condition of local and federal economies, which dictate the demand for electricity from coal-powered generating stations.

Surface Mining and Reclamation

Surface mining has a far more conspicuous impact upon the environment than does deep mining. However, surface mining allows the mineral to be more fully extracted when it is technically and economically feasible to remove the overlying nonmineral material (the *overburden*). In general, the economic feasibility of surface mining is determined by a ratio of the depth versus the value (quantity and quality) of the mineral. In the case of coal, this translates to a ratio, called the *stripping ratio*, between the depth to which overburden must be removed and the thickness of the coal seam. A stripping ratio of 10:1 is considered a rough upper limit, beyond which surface mining is no longer economical, although this generalization is often complicated by the presence of multiple coal seams or other extenuating circumstances.

Kinds of surface mines

Figure 1.3 illustrates the basic sequence of operations for the two most common types of surface mines: a strip mine and an open-pit mine. Other surface operations, such as contour mines and hilltop mines, are essentially adjustments of the above approaches to local topography.

Strip mining.   As Fig. 1.3A indicates, a strip mine is a long, narrow cut in the earth which progresses across the landscape, perpendicular to its long axis. Excavation, extraction, and reclamation are continuous, synchronized activities during strip mining, which is best for extensive, shallow deposits of coal, such as in the western United States.

The first step in strip mining involves the removal of vegetation and, in many cases, topsoil by bulldozers and self-loading scrapers. If topsoil is removed, it is redistributed on surfaces which have been backfilled and graded to an acceptable post-mining topography. Ideally, the topsoil is *live-handled*, which means that it is transported directly from the point of salvage to the surface where it will serve as a seedbed. However, topsoil must often be stockpiled between salvage and redistribution, during which time many of the desirable properties of the topsoil deteriorate.

The salvage and redistribution of topsoil is an important and controversial decision which depends upon the regulations and the reclamation objectives. The controversy revolves around the expense of the procedure and the degree to which the salvaged topsoil is the best pos-

A.

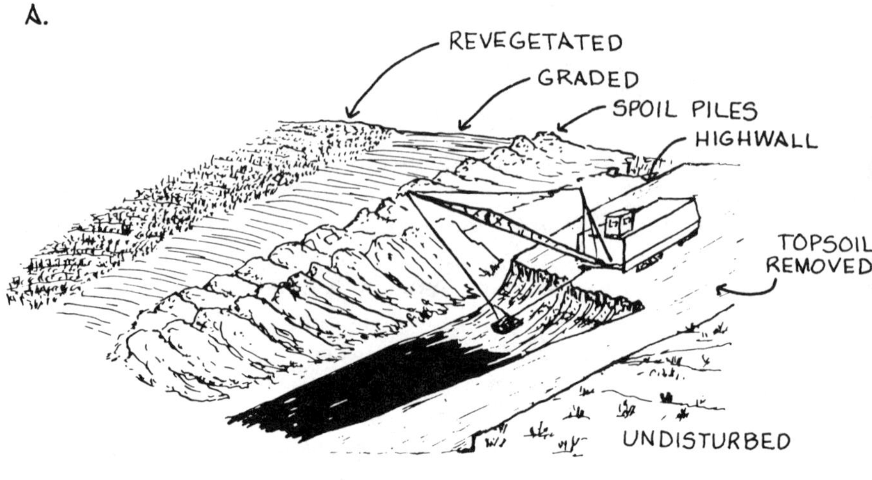

B.

Figure 1.3. Schematic illustrations of the operation of a strip mine (A) versus an open-pit mine (B).

sible medium for revegetation. This controversy is explored in many of the chapters which follow.

After the surface materials have been removed, the bedrock is drilled and packed with explosives. Detonations are carefully controlled to minimize release of airborne fragments and to maximize the shattering of the overburden. This shattered rock is removed by a steamshovel or dragline, and piled in cone-like banks of *spoil* alongside the cut.

The exposed coal is then shattered with explosives or fork-like rippers dragged behind a bulldozer. After the coal is removed, bulldozers push the spoil back over the trailing edge of the cut and grade the material into a flat or gently rolling topography. These earth-moving procedures *(back-filling* and *contouring)* are the most expensive part of reclamation. Meanwhile, the dragline continues removing overburden and exposing coal at the cliff-like *highwall*, which is the leading edge of the cut.

The next step in mine reclamation, and the first step in revegetation, is to construct a seedbed by distributing a 10- to 20-cm layer of topsoil over the graded spoils, or by altering the spoils themselves to an acceptable medium for seeding. Another option is to topdress the graded surface with a particularly favorable overburden stratum that has been *selectively handled*, that is, excavated and maintained for its desirable qualities as a revegetation medium. ("Selective handling" can also refer to the intentional isolation and deep burial of undesirable overburden strata, a procedure used for rocks with toxic concentrations of some elements.) The surface material may be physically altered by disking, plowing, and so forth, or it may be chemically altered by adding fertilizer, organic matter (including sludges), and other surface amendments.

Seed may be broadcast in a dry mix or in a wet suspension, or it may be drilled into the soil using a range-drill, which inserts each seed a few millimeters below the surface. Seeding may be followed by mulching with straw or other materials that will keep the seed and seedbed cool, dark, and moist. In windy climates, mulch may be crimped into the surface or attached with a tar-like tackifier. Trees and shrubs may be planted by hand, using horticultural procedures. These *containerized plantings* are preceded by a detailed regimen for germination and early growth in the greenhouse.

After seeding, the revegetated surface may receive periodic applications of irrigation, fertilizer, or additional seed. The revegetated stand must be maintained between five and ten years, depending on location, after which time the coverage, biomass, and productivity of the stand must compare favorably to the native vegetation in order to satisfy regulatory requirements.

Open-pit mining.   Open-pit mining (Fig. 1.3B), which is often used for heavy metals such as copper and uranium, and occasionally for coal, does not progress steadily across the landscape as does strip mining. Instead, a large (sometimes colossal) pit is excavated to expose the mineral, while the overburden is piled aside for the duration of the mining operation. The pit can be backfilled and the spoils reclaimed only after extraction is complete. Some mines consist of a sequence of small pits, and the reclamation program starts and stops depending on the rest of the operation. Other mines consist of a single pit which is too large to backfill. The immense quantities of spoil around the pit may be gradually graded and revegetated, although reclamation is largely voluntary for the heavy-metal (hardrock) industries which operate most megapit mines.

Tailings revegetation

Reclamation of spoils at hardrock mines is similar to the reclamation of spoils at coal mines. However, one aspect of the heavy-metal industry that presents a real challenge to reclamation is the production of great quantities of tailings at the mill sites. Tailings are fine rock particles from which ore has been extracted in the milling process. They are generally disposed in a network of dikes and settling ponds adjacent to the mill. One problem is the uniform texture of the tailings, which limits the availability of water to plants. Coarse tailings are like a pure sand soil, in that water infiltrates quickly, leaving a dry surface which is highly susceptible to wind erosion. Fine tailings are like a clay-rich soil, in that water may infiltrate very slowly and be retained on the particle surfaces so tightly as to be unavailable to plants.

Tailings may also have undesirable chemical properties. They may have levels of heavy metals which are potentially toxic to vegetation, herbivores, or soil organisms. Tailings may lack certain plant nutrients, or they may be extremely acid or alkaline, depending on the thermal and chemical procedures used to extract the ore. Finally, tailings from uranium milling may release radon gas or other radioactive materials.

Revegetation of mill tailings must address the adverse properties which appear to inhibit the germination and growth of plants and which otherwise deter reclamation. The role of science is to identify those adverse properties, to understand their inhibitory characteristics, and to develop remedial actions. This book does not specifically address tailings reclamation, but it presents research methods and reclamation procedures which may apply to the specific problems created by tailings. Likewise, the book's contents may selectively address reclamation problems associated with the residues of all forms of energy and mineral extraction, including oil shale, tar sands, and phosphate mining.

Introduction to the Chapters

So far, this introductory chapter has described the general objectives and organization of the book and its chapters. I have also attempted to provide sufficient background on the basic procedures and problems in reclamation, and on the environment of the San Juan Basin, to spare the contributors of that responsibility as they delve into their methods and case studies. The final goal of this chapter is to briefly introduce the individual contributions, thus enabling readers to preview or to review the various methods and perspectives.

Chapter 2:
Applications of geomorphology and
surface hydrology to reclamation.

Geologist Stephen G. Wells and ecologist Loren D. Potter describe factors which are important to the short- and long-term erodibility of native and reclaimed landscapes. One important application of geomorphology is to monitor and map the stability of landscapes prior to mining. A reclaimed surface in an unstable landscape will eventually be cut by major gullies regardless of the success of revegetation and stopgap measures to deter erosion.

Chapter 3:
Pre-mining assessments of
reclamation potential.

Plant Ecologist Loren D. Potter addresses the fundamental problem of pre-mining assessment: that effective land management requires reliable predictions of the ease and expense of reclamation prior to mining. These predictions of the behavior of the disturbed landscape are based on measurements and interpretations of the native landscape. The traditional approach has been inductive, wherein numerous environmental parameters (soil, climate, and so forth) are measured and interpreted. An easier, more reliable approach is deductive, wherein the native vegetation is observed to deduce what environmental factors predominate at the proposed mine site, which will therefore dictate the procedures and outcomes of reclamation.

Chapter 4:
The role of a government experiment station
and the experimental approach in developing
and testing reclamation practices.

Project director Earl F. Aldon describes how a U.S. Forest Service Experiment Station develops and tests combinations of procedures for reestablishing vegetation on disturbed land—procedures such as applying mulch, fertilizer, and other surface amendments. Aldon describes how experiments are conducted in a sequence which eliminates ineffective measures and thereby arrives at the best combination.

Chapter 5:
Understanding reclamation with models.

Environmental scientist Charles C. Reith explains the versatility of modeling as a tool for the organization and exploration of existing information on reclamation. Different types of models have been used to (1) simulate the behavior of reclaimed ecosystems, (2) identify relationships between environmental parameters and the outcomes of past reclamation, and (3) predict reclamation potential.

Chapter 6:
The selection of species for revegetation.

Agronomist William W. Fuller describes the rigorous procedures whereby species and varieties of plants are tested for their suitability for use in revegetation. Plant Materials Centers, operated by the Soil Conservation Service, identify candidate species for use in conservation practices, and test these plants for a wide variety of performance characteristics, including ease of germination, longevity, drouth tolerance, and self-propagation.

Chapter 7:
The study of soil water in reclamation.

Biologist David G. Scholl approaches reclamation from the perspective of water relations in the soil-plant-atmosphere continuum. The availability of water to vegetation, and the success of revegetation, depends on much more than the annual precipitation. Studies of water availability must consider the timing and intensity of precipitation, the loss of moisture to evaporation and percolation, and the physical and chemical nature of the soil.

Chapter 8:
The importance of the soil biota
and the ecosystem approach
in surface-mine reclamation.

Biologist Walter G. Whitford and civil engineer Ned Z. Elkins discuss the most neglected aspect of reclamation: the reestablishment of the soil biota. The ecosystem approach is used to describe the role of soil organisms in the cycling of nutrients and energy, and in the development of a true soil from mine spoil. The abundance, diversity, and influence of these organisms have been studied in laboratory microcosms and in mine-site research plots.

Chapter 9:
Vegetation ecology and the
determination of reclamation success.

Biologist Tim Fischer tackles the toughest question in reclamation: What is *successful* reclamation? Principles of plant ecology and statistics are used to assess the condition of the reclaimed plant community relative to the native vegetation or to other performance standards.

Acknowledgments

This book presents a wide variety of research methods and scientific findings in reclamation that would not have been possible without the invaluable support and cooperation of mining companies and granting agencies. The chapters do not specifically acknowledge the mine personnel who contributed to the implementation of studies, nor do they cite the contract numbers of the grants which provided support, because each chapter covers too many projects to address such specifics. However, the editors and authors would like to extend sincere appreciation to the following agencies and companies, who have been foremost in supporting our continued research in reclamation: U.S. Forest Service, Soil Conservation Service, New Mexico Energy and Minerals Department (Mining and Minerals Division), New Mexico Energy Institute, National Science Foundation, Office of Surface Mining, Bureau of Mines, Utah International Mining Company, Pittsburg and Midway Mining Company, Carbon Coal Company, Sun Belt Mining Company, and Anaconda Mining Company. The editors would also like to acknowledge the reliable assistance of Nancy Brennan-Reith, who did much of the typing and proofreading in this book. The drawings for Figs. 1.3, 2.5, 3.1, 3.3, 7.1, 7.2, and the front of the book were supplied by Loren Potter. All other drawings, except Figs. 8.4, 8.5, and 8.8, were done by Nancy Brennan-Reith.

References

Bieri, G. K. 1977. The environment and coal development in the San Juan Basin, New Mexico. *In* New Mexico Geological Society guidebook, 28th field conference, SJB III, 77-82.

Brady, N. C. 1974. The nature and properties of soils. New York, NY: MacMillan Publishing Co.

Fassett, J. E., and J. S. Hinds. 1971. Geology and fuel resources of the Fruitland Formation and Kirtland Shale of the San Juan Basin, New Mexico and Colorado. U.S. Geological Survey professional paper 676.

Gould, W. J., D. Rai, and P. J. Wierenga. 1975. Problems in reclamation of coal mine spoils in New mexico. *In* M.K. Wali, ed., Practices and problems of land-reclamation in western North America, 107-126. Grand Forks, ND: University of North Dakota Press.

Grant, P. R. 1981. Overview of energy resources in New Mexico. *In* Environmental hydrology and geology in New Mexico, 15-20. New Mexico Geological Society special publication no. 10.

2

Applications of Geomorphology to Reclamation

Stephen G. Wells and Loren D. Potter

Introduction

The first major step in surface mining is to create a drastic change in the geomorphology of the landscape. The natural stratification of bedrock, regolith, and soils is changed, as is the natural system of drainage. Rock layers which were originally horizontal are shattered and mixed in irregular piles of spoil.

It is impossible for reclamation to reconstruct the original geologic structure, but it is imperative that spoils be recontoured to a topography that will be relatively *stable* against processes of erosion such as slumping, gullying, rilling, piping, and sheetwash. In fact, a primary goal of reclamation, as dictated by the Surface Mining Control and Reclamation Act (SMCRA), is to restore the landscape to a stable configuration. However, the attainment of stability is a complex challenge that depends upon the interaction of many of the variables which will be mentioned in this chapter. Still another challenge in reclamation is to restore a drainage system that will drain the reclaimed spoils and anastomose into the drainage network of the surrounding landscape.

Geomorphology is the study of the shape of the land and the arrangement of bedrock, soils, and regolith (the mantle of weathered parent material). It is also the study of the *processes* which are responsible for these attributes of a landscape. Considering the impact of surface mining upon the appearance and behavior of a landscape, it is obvious that the geomorphic approach has much to offer in the study of reclamation. This chapter discusses the geomorphic approach to reclamation, which has the following principal objectives: (1) to describe the evolution of landscapes, (2) to quantify the relationships between landform processes, (3) to measure rates of surficial processes, and (4) to apply the results to the reclamation of disturbed lands.

Geomorphic concepts and reclamation regulations

Geomorphic evaluation of the success of reclamation is usually accomplished by measuring selected landscape conditions and responses prior to mining and after reclamation. Another option is to use a landscape similar to the mined landscape as a standard for comparison during the reclamation period. Two frequently used measures of success are the nature of the vegetation cover and the stability of the area against erosion, usually measured as sediment yield. Coal mines in the arid and semiarid West often occur where climate and vegetation combine to maximize sediment yield, as is illustrated by the relationship between effective precipitation, vegetation, and sediment yield (Fig. 2.1). This chapter addresses sedimentation and other geomorphic processes as measures of reclamation success. These relationships among geomorphology, hydrology, mining, and reclamation are elaborated in greater technical detail in a report by Wells (1982) on the San Juan Basin.

Regulations in SMCRA restrict the amount of sediment which can exit a mine property via natural drainages to tens of milligrams per liter (mg/l). This makes erosion control one of the greatest reclamation problems facing coal-mining industries (Stiller et al. 1980). The irony of such regulations is that most natural streams in arid and semiarid western environments have extremely high concentrations of suspended sediment, usually measured in thousands or tens of thousands of milligrams per liter—three orders of magnitude greater than what is expected from reclaimed watersheds. A similar situation exists with regard to total solute concentrations.

Sediment yields are relatively high in arid and semiarid drainage basins because of the sparseness of vegetation. Sediment release is still greater in situations such as the San Juan Basin, where rainfall comes infrequently but torrentially, generating great volumes of runoff. Most arid and semiarid drainage systems are adjusted to carrying large quantities of sediment during runoff events (Schumm 1977). Thus, if the mine operator conforms to federal regulations by limiting the sediment load draining off the mine area, the natural drainageways may suddenly be changed. Stream channels that are adjusted or stable under naturally high sediment loads become unstable if significant alterations occur. The effect of such reductions on the long-term stability of drainageways remains unknown.

As mentioned above, SMCRA requires that erosion be controlled by

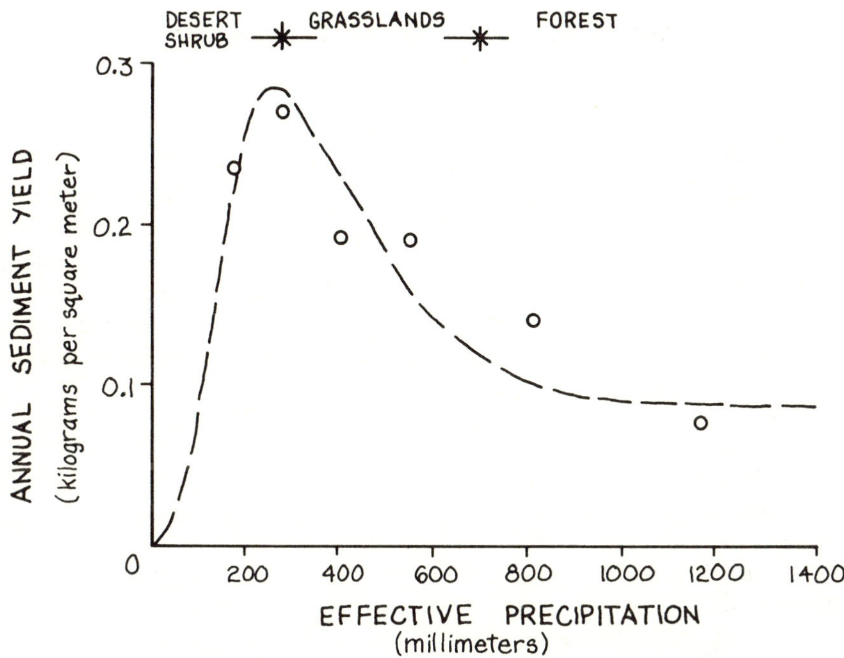

Figure 2.1. Relationship between effective precipitation (mean annual), vegetation, and sediment yield (from a given surface area) in the western U.S. (Modified from Langbein and Schumm, 1958.)

restoring post-mining landscapes to stable configurations. However, landscape stability is a function of many variables, including climate, bedrock, geology, and geomorphic history. The above examples of sediment and solute regulations indicate that governmental restrictions may not be compatible with the behavior of these variables on natural and reclaimed landscapes. To address these potential difficulties, the reclamation specialist must understand geomorphic processes as completely as possible. Baseline studies are described below which illustrate the kinds of measurements that are needed for a complete understanding. These studies are expressed with reference to the San Juan Basin, which serves as a discrete geologic region for the measurement and characterization of geomorphic variability.

Geomorphological Pre-mining Assessment

Regional landscape types

The first step in studying the geomorphology of a region is to identify and characterize the regional landscape types. The coal-bearing formations in the San Juan Basin have landscapes of varying ages, which we divided into four major types: (1) upland surfaces including pediment and terrace remnants capped with upper Quaternary (Ice Age) deposits; (2) modern, alluvial valley floors and active channels; (3) eolian (windblown) landforms and deposits; and (4) badland topography developed on exposed Cretaceous bedrock.

These landscape types differ with respect to geomorphic history, character of surficial material, and types and rates of surficial processes. An area to be mined may be composed of a mixture of any number of types, each with its own features of stability. However, within a given area, the surficial processes interact between landscapes.

Upland surfaces.   Uplands consist of the flat surfaces of horizontal strata or the eroded edges of outcrops covered with a mantle of Pliocene to Holocene (Recent) deposits. Some of the best preserved mantle deposits overlay shales and mudstones of the Fruitland and Kirtland formations (Schultz 1980). Upland surfaces are at several different topographic levels. These most stable landscapes are gently sloping, well-vegetated mesas with broad, unconnected drainage lines.

Valley floors.   Valley floors are composed of active channels flanked by broad alluvial plains. Sediments of the plains are deposited by alluvial fans, sheetwash, and eolian processes. Most valley floors are low terraces which are well vegetated, and may be deeply dissected (Gutierrez 1980). These are storage areas for sediments from tributaries and badland regions.

Eolian landforms and deposits.   Eolian deposits consist of dunes of a variety of shapes and sizes. These are both stabilized and active, with the latter usually found on valley floors. Some stabilized dunes have well-developed soil horizons reflecting their antiquity.

Badland topography.   Badland topography is found on steep hillslopes that are transitional between the uplands and the valley floors. Erosion to the bedrock results in high, local relief with many drainages per unit area. The majority of the badlands are recent in origin.

Quantification of landscape stability

The next step is to characterize these landscape types in terms of their stability, which has been defined by Thornes (1979) as a landform's sen-

sitivity to change. Landscape stability is generally assessed in terms of
river processes (Schumm 1977), pedogenic processes (Gile et al. 1981)
or mass-movement processes, such as landslides (Godfrey 1978). For reclamation purposes, a more practical definition of stability considers all
regional landscape processes. However, the difficulties of measuring so
many processes has prevented such characterizations of regional stability on a large scale. Wells and Rose (1981) measured a wide variety of
geomorphic processes on small, individual landscape components, each
of which was discrete and regional. Results of these kinds of studies
can then be extrapolated to larger areas.

Relative stability can be determined for all of the landscape components over an entire strippable coal belt. As mining and reclamation
progress across an area, measurements of areal extent and relative stability can be compared to the entire pre-mining landscape, or they can
be compared to specific pre- and post-mining components of similar morphology. It is important to realize that all landforms fluctuate somewhat about a mean trend, but such fluctuations are accepted as typical
of a steady-state condition (Thornes 1979). Thus, in applied (reclamation) landscape-stability studies, minor fluctuations around a trend
should not be interpreted as failures or instability, but as part of an acceptable dynamic stability. Instability would be indicated by *long-term* trends
and *major changes* in trends. The same concepts apply to the succession of plant communities and to the maintenance of a dynamic equilibrium of climax vegetation.

A difficult landscape situation to evaluate arises when no trends are
evident, but the landform has the *potential* for instability. Threshold
conditions may exist beyond which internally or externally applied forces trigger rapid landform changes. This is analogous to suddenly reaching the critical point in titration of a basic solution with an acid. An
example of an external force which may exceed the threshold of stability might be a summer thunderstorm of sufficient intensity to generate
runoff and erosion. An example of an internal threshold is indicated in
Fig. 2.2, where an alluvial valley floor may be incised by gullies if its
slope is too steep for the surface area of the watershed. Gully incision
is considered an unstable condition because it results in large volumes
of sediment. Instability associated with gullies occurs frequently in undisturbed landscapes in the San Juan Basin.

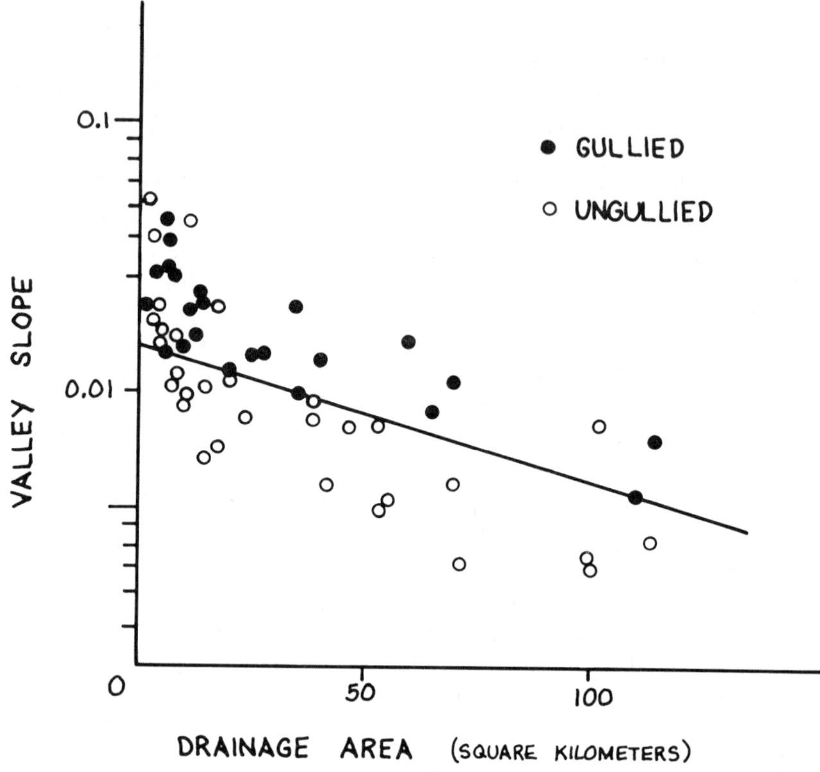

Figure 2.2.   Gully-incision threshold for a given valley slope and drainage area. Line indicates threshold conditions. (Modified from Patton and Schumm 1975.)

Surficial-process units

Most future mining for coal in the San Juan Basin will occur in coal-bearing areas of the Cretaceous Fruitland Formation and Mesa Verde Group sedimentary rocks of the Basin. Thus, analyses of stability must emphasize those geologic units. The most valuable information for predicting the stability of reclaimed surfaces in these units is baseline data describing their natural surficial processes and geomorphic conditions. Unfortunately, such data are scarce (Bierei 1977).

Our strategy was to use field measurements and stereo aerial photographs to subdivide the aforementioned landscape types into mappable units based on surficial processes. Tables 2.1 and 2.2 list seven units which reflect the modern-day processes acting on deposits and bedrock

Table 2.1. Geologic, morphologic and vegetational properties of surficial-process units[*]

Surficial-Process Units	Landforms	Type of Surficial Material	Age of Surficial Materials[1]	Range in Slope Gradient (percent)	Bare (unvegetated) Soil[2] (percent)
QTp(s)	pediment and terrace remnants	fluvial clay, silt, sand, and gravel, usually indurated	Pliocene(?) to Pleistocene	0–1	45–64
	stabilized dunes and sand sheets	eolian sands	late Pleistocene to Holocene	0–1	41–58
QTp(t)	dissected QTp(s)	includes some material reworked from QTp(s)	same as QTp(s)	12–28	85
Qal_3	broad vegetated channels	fluvial clay, silt, and sand	middle to late Holocene	1–4	41–58
KTbb	badlands	thin clayey to silty weathered mantle	late Holocene to modern	10–40	94
Qal_2	low terraces, alluvial fans, and stabilized dunes	fluvial clay, silt and sand; eolian sand	late Holocene	1–5	35–58
Qal_1	active channels	fluvial sand and gravel	present day	0.3–1	90–100
Qds	active dunes	eolian sand	late Holocene	1–10	65–100

[*]See Figure 2.3 for examples of the spatial distribution of these units.

[1]Data taken from Schultz (1980) and Wells and Gutierrez (1981).

[2]Data taken from U.S. Department of the Interior, Bureau of Land Management (1976), Figure 33.

Source: Wells and Rose 1981.

Table 2.2. Characteristics of major surficial-process units in the strippable-coal belt north of the Chaco River.[*]

Surficial-Process Units	Landscape Type	Dominant Geomorphic Process(es)	Erosion Type (vertical downcutting)	Rate (cm/yr)	Infiltration[1,5] Rate (cm/hr)	Detachability[1,6] Rate (kg/m²/hr)	Estimated Sediment Yield[3,7] (cm³/m²/yr)	Measured Sediment Yield[4,8] (cm³/m²/yr)
QTp(s)	stable upland surface	erosional eolian stability, deposition	sheetwash	(undetermined)	>20	100–1,000	0–24	(undetermined)
QTp(t)	transition between upland surface and badland	slope erosion	rilling and sheetwash	0.3–2.0	1–20	10–1,000	810–1,500	2,800–20,000
$Qal_3{}^2$	upland surface	fluvial erosion and deposition	gully incision and sheetwash	—	—	—	—	—
Ktbb	badland	slope erosion	creep and sheetwash	0.1–2.0	0–1	1–10	570–1,500	950–26,000
Qal_2	valley floor	slope erosion and fluvial deposition (0.1–0.5 cm/yr)	sheetwash and gully incision	—	1–10	100–1,000	120–310	3,600–18,000
$Qal_1{}^2$	valley floor	fluvial transport	—	—	—	—	430–570	—

| Qd_4 | valley floor and upland surface (not differentiated on upland surfaces) | eolian deposition | — | — | 20 | 100–1,000 | up to 190 (fluvial) | — |

*See Figure 2.3 for examples of the special distribution of these units.

[1]Data taken from U.S. Department of the Interior, Bureau of Land Management (1976).

[2]Data not available on process rates and yields.

[3]Determined by the unmodified Pacific Southwest Inter-Agency Committee (PSIAC) method; data taken from U.S. Department of the Interior, Bureau of Land Management (1976).

[4]Taken from Wells (1978) and Gutierrez (1980).

[5]Rate of water that is absorbed into ground using a single-ring constant infiltrometer.

[6]Susceptibility of a soil to erosion by running water determined under controlled laboratory conditions.

[7]Empirical (PSIAC) method of determining volme of sediment yielded from a land surface under assumed precipitation conditions.

[8]Determined in the field during precipitation events.

Source: Wells and Rose 1980.

of various ages. These units have been mapped for some of the strippable coal belt in the San Juan Basin on U.S. Geological Survey (USGS) 7.5-minute topographic quadrangle maps; Figure 2.3 illustrates a representative map. Much of the area in Fig. 2.3 consists of QTp(t), which includes dissected pediments, terraces, and stabilized sands, and KTbb, which is badland topography. These are the two least stable units. The most stable unit is the QTp(s), a stable upland surface which produces little runoff or sediment. Extensive areas of Qal_2 (low terraces or alluvial fans) are potentially unstable. In spite of moderately high infiltration rates and good vegetational cover, numerous gullies and stream channels have dissected this unit, producing higher than expected sediment yields.

When the area to be mined is superimposed on a map like Fig. 2.3, which shows relationships of stability to drainage systems, the reclamationist can better predict the success of reclamation and anticipate potential hazards from adjacent unstable areas. It is also an aid in designing the restored drainage system.

The geomorphic history is important in determining stability. The stable upland units, QTp(s), result from a mantle of Pleistocene and Holocene eolian sand. This allows for maximum infiltration, good root penetration, favorable moisture and aeration for plant growth, and maximum vegetational cover. Erosional instability is due principally to the lowering of base level over time. Upland surfaces are dissected by a process called *base-level lowering*, which describes how landscapes are cut down to the level of major streambeds. Lowering of base level strips the protective mantle off these surfaces and exposes underlying bedrock. Hillslope erosion then creates a badland topography, much of which has developed within the last several thousand years. These areas have high local relief, low infiltration rates, and rapid development of weathered mantles, all of which combine to generate high sediment yields.

Measurement of Surficial Processes

Processes and variables

Table 2.3 lists the ways in which the surficial processes for each mappable unit may be measured. Instrumentation of such units may require some reduction into subbasins. Within each watershed, a description must be obtained of the surficial deposits, weathered mantles, and bedrock lithologies. Descriptions are based upon analyses of clay mineralogy, major soluble cations, textural parameters, and grain-size distribution. These analyses help identify local sources of sediments and,

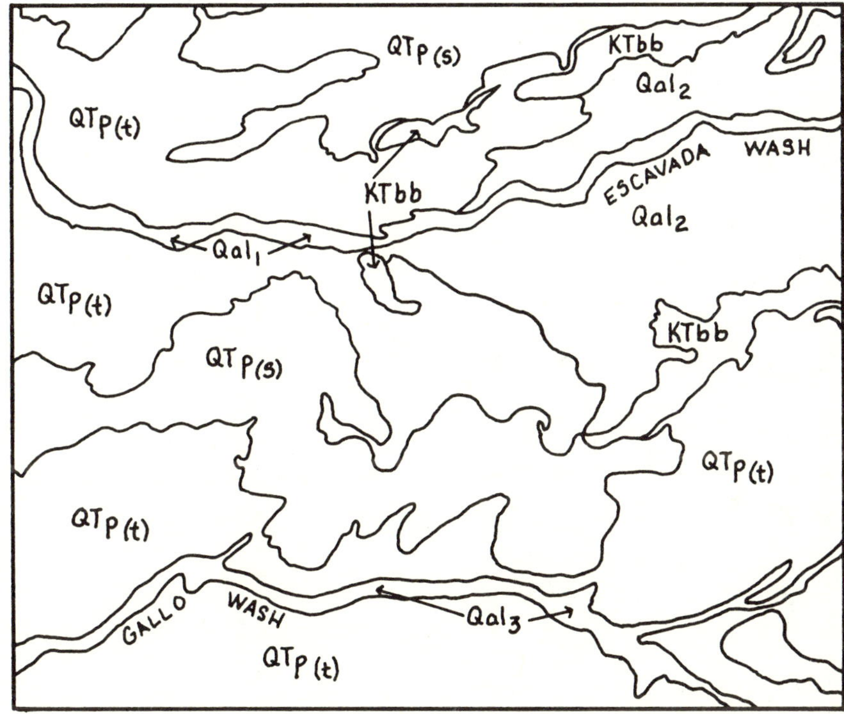

Figure 2.3. Distribution of surficial-process units in part of the strip-pable-coal area of the Fruitland Formation of the San Juan Basin (from Gutierrez 1980). See Tables 2.1 and 2.2 for explanation of map units.

thus, the location of major erosional processes and the transport paths of sediments.

Figure 2.4 describes variables affecting surficial processes according to their temporal and spatial scope. Variables are also classified according to the source of their influence upon landscape stability. Of the surficial processes in Fig. 2.4, those which are measured are ones depending upon the variables described as the Class-A magnitude of relatively recent time and small area. External variables include precipitation characteristics, alluvial-fill geometry, gully headcutting rates, and bedrock type and continuity. These variables influence such processes as *rilling, gullying, headcutting,* and *piping.* Internal variables include infiltration conditions; composition, geometry, and surface roughness of hillslopes; and drainage density (ratio of the total length of all channels to area of basin)

Table 2.3. Summary of techniques used to characterize the geomorphic variables.

Techniques	Geomorphic Variables									
	Surficial Deposits	Climate	Vegetation Type and Density	Relief	Hydrology (Runoff and Sediment Yield per Unit Area)	Drainage Network Morphology	Hillslope Morphology	Hydrology (Total Runoff and Sediment Yield)	Valley-Fill Morphology	Depositional Features of Sediment-Settling Ponds
Field Mapping	X									
Precipitation Stations		X								
Vegetation Grids			X							
Topographic Surveying				X			X		X	X
Infiltration Tests					X					
Erosion-Pin Transects					X					X
Particle-Size Analysis					X		X	X	X	X
Clay-Mineral Identification					X		X	X	X	X
Morphometry						X				
Surface Roughness							X			
Gauging Station								X		

Source: Modified from Jercinovic 1984.

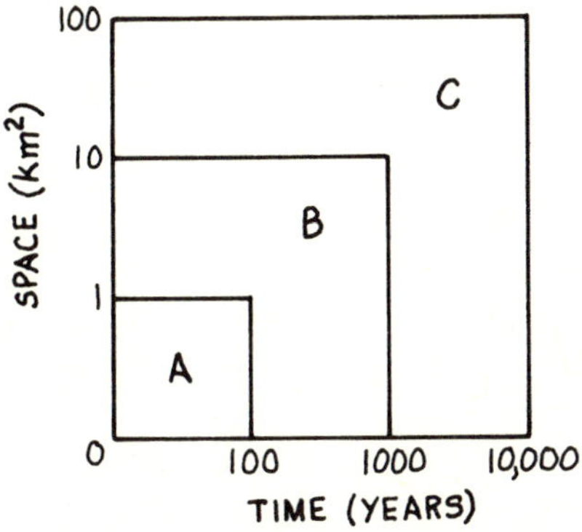

VARIABLES AFFECTING LANDSCAPE STABILITY		
MAGNITUDE CLASSES OF AREA AND TIME	INTERNAL	EXTERNAL
A	• HILLSLOPE GEOMETRY, GRADIENT & COMPOSITION • SURFACE ROUGHNESS • HILLSLOPE DRAINAGE DENSITY & FREQUENCY • SURFACE & SUBSURFACE BIOMASS • INFILTRATION CONDITIONS • SEDIMENT YIELD & RUNOFF PER UNIT AREA • VEGETATION DENSITY	• RATE OF ARROYO HEADCUTTING • VEGETATION DENSITY • SEDIMENT YIELD & RUNOFF (TOTAL) • VALLEY-FILL GEOMETRY • BEDROCK HETEROGENEITY • PRECIPITATION INTENSITY, DURATION & FREQUENCY • DRAINAGE DENSITY & FREQUENCY (WATERSHED)
B	• HILLSLOPE RELIEF & GRADIENT • HILLSLOPE NETWORK • RATES OF WEATHERING & SOIL DEVELOPMENT • DEGRADATION & AGGRADATION RATES	• VEGETATION DENSITY • VALLEY SLOPE • DRAINAGE AREA • CHANNEL PATTERN & GEOMETRY • WATERSHED DRAINAGE NETWORK • BEDROCK GEOMETRY • CLIMATIC PERTURBATIONS
C	• VEGETATION DENSITY • DEGREE OF SOIL DEVELOPMENT • DEGRADATION & AGGRADATION RATES	• VEGETATION DENSITY • BASE-LEVEL CHANGE • CLIMATIC CHANGE • BEDROCK GEOLOGY

Figure 2.4. Internal and external variables affecting reclaimed and undisturbed landscape stability over three magnitudes of time and area. Magnitudes of classes, A, B, and C variables, given in upper portion of diagram.

and configuration. These variables influence *soil creep, mudflows, debris-flows, sheetflow,* and *rilling.*

The valley-fill landscape component is dominated by *gully incision processes* and is therefore influenced by variables which are external rather than internal. The disturbed bedrock, undisturbed bedrock, and reclaimed bedrock landscape components are characterized by sheetflow, soil creep, and rilling, and are more sensitive to internal rather than external variables.

Figure 2.5 is a schematic representation of the runoff processes which occur on badland slopes and reclaimed surfaces with similar physical and chemical characteristics. On shale surfaces in the San Juan Basin, the initial sealing of the surface takes place very rapidly (ten to twenty-five minutes) after precipitation begins. Up to an hour later, small surficial mudflows are initiated and overland flow begins. Small surficial (<2 cm) mudflows are followed by a period of relatively clean sheetflow. But, if precipitation persists, larger mud and debris flows will occur within two to four hours, creating incised scars and lobate or fan-like deposits. The flow results from saturation of the mantle, the impermeability of the shale bedrock, the increased pore pressure, and the lowering of the coefficient of friction.

Measurement techniques

Table 2.3 summarizes techniques which can be used to study variables which are needed in applications of geomorphology to reclamation. Many of these techniques were used in studies of geomorphology and reclamation in the San Juan Basin, for which Jercinovic (1984) provides extensive details on the specific methodologies and data derived therefrom.

Precipitation. The seasonality and intensity of precipitation is closely related to surficial processes. In the San Juan Basin, winter precipitation occurs as rain and snow of low intensity which lasts for several days. In contrast, summer precipitation occurs as short-duration, high-intensity rain from convective thunderstorms. These events are best measured with recording rain gauges, to include intensity; but a network of cheap, nonrecording rain gauges may provide valuable data, especially concerning the sporadic areal distribution of precipitation patterns.

Evaporation. Evaporation can be measured with relatively simple open-pan evaporimeters.

Changes in relief. Erosion-pin transects are frequently placed at different locations within a basin: (1) headwaters, (2) midbasin, and (3) basin mouth. Relief changes may be due to a reduction of elevation because of surface erosion or to an increase of elevation because of the deposi-

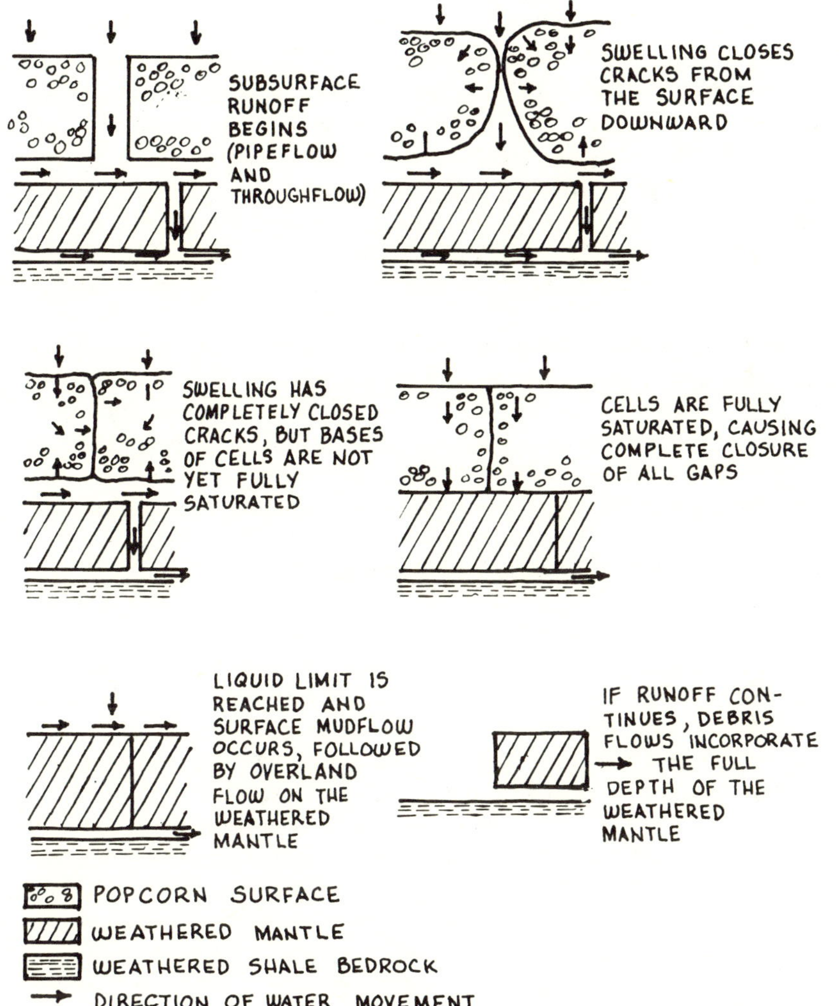

Figure 2.5. Schematic representation of runoff processes on badland hillslopes. (Modified from Hogg 1978.)

tion of detritus from upslope. Erosion-pin transects can also detect changes in elevation due to frost heaving and clay swelling.

A second method of measuring rates and types of surficial processes is the erosion grid. Four stakes driven into the ground serve as permanent monuments. A one-meter square grid containing twenty-five control points is placed over the four stakes. Detailed microtopographic maps are made by using a vertical rod to measure the height between ground surface and grid frame at each of the twenty-five control points. The frame is then removed until the next measurement is made. Results of the erosion grid are interpreted by a computer to produce topographic maps, which may be compared to detect changes in ground-surface configuration. The differences in elevations can then be plotted and contoured by computer to indicate areas and amounts of erosion or deposition.

Sediment and runoff.   Surface processes may also be measured by sediment traps. Sediment dams trap runoff and allow the settling out of sediment, which can then be measured. Sediment barrels can be installed at the base of drainageways and are relatively inexpensive to construct with PVC pipe and used 55-gallon drums.

Channel runoff can be measured by preprogrammed ultrasonic level and flow recorders at Parshall flumes. Sedimentologic budgets can be measured by automatic suspended-sediment samplers, which are electronically tied into the flow devices to sample on a flow-proportional basis. The above equipment, along with H-flumes, Stevens water-level recorders, and single-stage sediment samplers, were used to measure the levels and characteristics of runoff from native and reclaimed landscapes in the San Juan Basin (Wells and Rose 1981).

Our experience with the full range of sophisticated intrumentation has confirmed a long-standing belief that the latest technology is not always the best technology for obtaining useful scientific data. The high cost limits the number of instruments and thereby restricts sample size. Furthermore, complicated instruments often fail, resulting in a loss of critical data. The use of simple, inexpensive field equipment allows for large grids and better sample sizes, often giving a more complete picture of the scientific phenomenon under study.

Relationship of Geomorphic Processes to Reclamation

Interactions between landscape components

More complex than estimating the effect of disturbance and reclamation on a single component of the landscape is the understanding of the interactions between components. Because the entire watershed has developed as a coordinated system, a disturbance such as mining and reclamation may have effects upslope, downslope, and laterally. The effects may be especially complex when several mines occur on the same watershed. Areas not directly disturbed by mining may become unstable until the entire system has had time to adjust. Conversely, reclamationists must look out for disturbances which are unrelated to mining and which occur outside the reclaimed areas, or even outside the mine lease. Disturbances in this category can adversely influence reclamation regardless of the procedures used for revegetation and stabilization. A common example in the arid West is the incision of valley floors by gullies which migrate upvalley by headcutting.

Rates of headward migration of gullies in New Mexico have been measured in meters per year (Malde and Scott 1977, Mills and Gardner 1983). Bank cutting is accelerated by piping, which removes soil by subsurface erosion. Surface flow is funneled into subsurface conduits as large as 5 m in diameter (in badland areas). Piping is common in areas of shale which is overlain by a veneer of sandy alluvium, or in weathered bedrock of high permeability, vertical structure, and high sodium content. Water draining into steep-walled channels develops a powerful hydraulic head (water pressure) which creates and enlarges pipes. Eventually, these pipes integrate and collapse to form gullies.

In the Rio Puerco drainage, which is adjacent to the southeast corner of the San Juan Basin, we observed lateral extensions of pipe-gully systems averaging 6 m/year (Wells et al. 1983). Such gullying may endanger the stability of a reclaimed area. This danger points out the necessity for proper management of critical areas outside the area of actual mining operation.

Variables pertaining to reclamation

As mentioned earlier, and listed in Fig. 2.4, the variables responsible for the surficial processes involved in landscape stability can be divided into two types (Wells and Rose 1981).

Internal variables.   Internal variables are related to surface and subsurface conditions within reclaimed areas. These are initially influenced by engineering practices in the preparation of spoil areas for revegetation.

External variables.   External variables are related to all conditions geographically outside the reclaimed areas. These factors, which include climatic conditions, bedrock geology, and geomorphic history, can rarely be manipulated, but the extent of their influence should be understood in reclamation planning.

Scope of influence.   Figure 2.4 also classifies variables according to their scope in time and space. The number of variables which are important to reclamation decreases as the scale in space or time increases. Understandably, reclamation practices emphasize short-term, small-scale processes. However, longer-time, larger-scale variables (both internal and external) may be especially important, and should be considered in applications of geomorphology to reclamation. Because of their significance, these variables have also been measured and related to reclamation, as will be discussed in the next section.

Examples of the use of landscape components in reclamation

The relation of mining and reclamation to geomorphology will be illustrated by discussing the use of major surficial-process units of the San Juan Basin (Tables 2.1 and 2.2).

QTp(s) unit.   This is the most stable landscape form. Extremely low erosion rates and sediment yields on reclaimed surfaces could be obtained by establishing the following characteristics in the reclamation design: (1) slope gradients of 1% or less; (2) permeable topdressing material composed of eolian sands and sandy deposits of ancient fluvial systems; (3) broad, shallow drainages (Qal_3); and (4) a vegetational cover which is comparable to the native condition. The porous surface materials favor infiltration, thus decreasing runoff and sediment removal. This is especially important where spoils consist of shales with high concentrations of sodium (1,000 mg/l or higher). The favorable QTp(s) deposits vary in thickness to as deep as 10 m.

QTp(t) and KTbb units.   These are the least stable landforms. The former consists of the transitional dissected slopes between uplands and the badlands of the KTbb unit. Figure 2.6 illustrates the latter unit in a natural San Juan Basin landscape. The principal undesirable characteristics of such a landscape, which should be avoided or reduced in the reclamation design include: (1) moderate to low infiltration rates, (2) high detachability rates with extensive (>80%) bare ground, (3) topsoil material with high clay content, and (4) slope gradients in excess of 10%. These units generate the highest volume of runoff and sediment, which contains high concentrations of soluble cations. The critical time affecting the stability of these units is the summer period of intense thunderstorms with precipitation in excess of 1 cm/hr (Fig. 2.7). Where the KTbb unit is more common than the upland QTp(s), there will be less surface sandy soil available for topdressing, a situation which may be remedied by plating the surface with sandstone spoils material.

The Qal_1, Qal_2, and Qal_3 units.   These are areas where high volumes of sediment are transported and stored. The active stream channels (Qal_1) are either braided or meandering and are adjusted to high sediment loads

Figure 2.6. Area of deposition (Qal$_2$) at base of badlands (KTbb) in a strippable-coal belt of the Fruitland Formation. Active channels are the light-toned areas at the base of the badlands. (From Wells and Rose 1981.)

from KTbb areas. The Qal$_2$ unit consists of low terraces, alluvial fans, and stabilized dunes where sediments are deposited and stored because of the low gradients and extensive vegetative cover. The Qal$_3$ unit consists of broad vegetated upper terraces. The latter two units are metastable (intermediate in stability), exhibiting both deposition and incision. The steady-state condition of the drainage system depends on a systematic relationship between the length of the stream per unit area (drainage density) and the number of stream segments per unit area (drainage frequency). If these areas were to be mined, this ratio must be restored to achieve long-term stability for the channel system.

Geomorphic study of a specific mine site

We used a variety of instruments on watersheds at Mine C to compare geomorphic process between undisturbed (pre-mining), disturbed (active-mining), and reclaimed (post-mining) landscapes (Jercinovic 1984). During a short-term storm event, the reclaimed and undisturbed watersheds responded similarly in runoff and infiltration. The disturbed water-

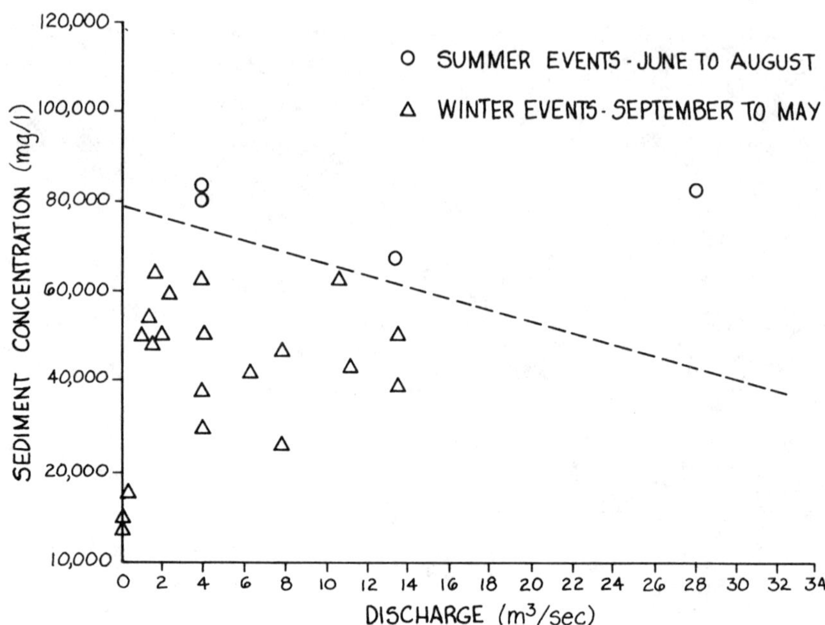

Figure 2.7. Seasonal precipitation patterns and associated sediment concentrations in the strippable-coal belt of the San Juan Basin (from Gutierrez 1980). Broken line separates seasonal events.

shed had 32 times as much runoff and 1.5 times less infiltration per unit area than either of the other watersheds. The reclaimed hillslopes have extremely rough surfaces composed of unsorted spoils, which have been furrowed along the contour and seeded with perennial bunch grasses. The unsorted spoils create a favorable, irregular microtopography characterized by protruding boulders and very small subsidence features (slumps and pits). These conditions increase infiltration, resulting in lower rates of runoff and erosion than has been reported for other western reclaimed areas.

The success here is due to a combination of internal geomorphic variables. In the undisturbed watershed, the high infiltration/low runoff is attributed to both the external and internal geomorphic variables. Bedrock type and continuity (external variables) produce the equivalent of a flattop mesa with a sandy regolith (mantle of loose rock material). This regolith, in turn, provides permeable conditions and promotes surface roughness (internal variables) by supporting dense vegetation. The disturbed watershed was in a temporary condition associated with steep-sided piles of spoil. The watershed had low infiltration and high runoff in response to the internal variables of hillslope geometry and the lack of vegetational cover. The effect of the increased sediment discharge was lessened by channel adjustments in the stream network. The temporary storage of sediment in the stream buffered the remainder of the watershed downstream from the disturbed landscape component.

All inherently stable hillslopes, whether natural or reclaimed, may be transformed into an unstable condition. This process occurs naturally where watersheds have an unstable valley-fill landscape component. Active headcutting and gully incision naturally progress upslope, cutting into hillslope components. In the vicinity of Mine C, this natural process is hard to stop and confine to the valley-fill. In both reclaimed and undisturbed watersheds, most sediment was from headcutting and bank slumping, not from erosion of the hillslopes. Therefore, in the area of Mine C, and presumably elsewhere in the San Juan Basin, external variables such as gully processes pose the greatest threat to the reclaimed landscape component and ultimately the success of post-mining reclamation.

Summary

A major goal of reclamation is to reduce the amount of erosion and sediment yield. However, regulatory agencies must recognize that these semiarid watersheds of New Mexico produce some of the highest sediment yields in the western United States. Mining operations should not be allowed to seriously increase the amount of erosion and sedimentation of the complex of landscape components, but neither should they be expected to reduce sedimentation levels to that of a densely vegetated area. Other aspects of regulatory policy make reference to "restoration to conditions comparable to those before mining." If those same standards were used in regard to runoff and sedimentation, mining and reclamation programs should be able to meet pre-mining conditions and even improve them.

Both for a prediction and measure of reclamation success, each landscape component of the area should be measured for its relative stability. This is a measure of: (1) rate of surficial processes (erosion), (2) the potential for instability (threshold conditions), (3) the complex responses of the geomorphic system (interaction of landscape components to a disturbance), and (4) the significant variables which affect stability at different time periods and in different areas. Reclamation procedures should include evaluations of internal variables (surface and subsurface conditions of reclaimed areas) and external variables (conditions outside the reclaimed areas). Properly engineered and revegetated areas may be adversely affected by external variables such as rates of gully incision or headcutting. This suggests that more attention be given to the management of critical areas outside the reclaimed area, both for the long-term preservation of the stability of the reclaimed area and the total landscape.

In the strippable coal areas of the Basin seven types of regional surficial-process units have been defined, mapped, and designated as to relative stability. These units serve as analogs for designing reclaimed landscapes. Those characteristics of the most stable landscape should be used as guidelines for designing the reclamation program. These characteristics include: (1) slope gradients of 1% or less; (2) permeable topdressing material composed of eolian sands and sandy fluvial deposits, or equivalents; (3) broad, low-relief drainage lines which have low gradients; and (4) vegetation cover equivalent to the pre-mining condition.

Initial studies at Mine C during a three-day rainstorm event indicated that the reclaimed watersheds were equal to the undisturbed watersheds in low runoff because of the permeability of reclaimed or natural surfaces. The high sediment yield in both undisturbed and reclaimed watersheds came from gully banks and not from the altered or natural hillslopes. The stability of reclaimed watersheds in this area is more susceptible to external variables than to the design and practices of reclamation.

More research is needed to fully appreciate the interactions of the landscape components. It will then be possible to better predict both the effect of the reclaimed area on the surrounding landscape and the rate and effect of the external variables on the reclaimed site. This recognizes the need to manage the entire system and not just a single component.

Much of the research reported here was sponsored by the New Mexico Energy and Minerals Department, Grant No. 68-3111 and by the Pittsburg and Midway Coal Company. We express thanks to the sponsors, to Dr. C. Nathan, New Mexico Energy Institute, Socorro, New Mexico, and to many individuals for their field work and lab analyses, especially to Devon (Rose) Jercinovic.

References

Bierei, G. R. 1977. The environment and coal development in the San Juan Basin, New Mexico. *In* New Mexico Geological Society guidebook 28th field conference, SJB III, 77-82.

Gile, L. H., J. W. Hawley, and R. B. Grossman. 1981. Soils and geomorphology in a basin and range area of southern New Mexico, Guidebook to the desert project. Socorro, NM: New Mexico Bureau of Mines and Mineral Resources memoir 139. In press.

Godfrey, A. E. 1978. Land surface instability on the Wasatch Plateau, central Utah. Utah Geology 5:131-141.

Gutierrez, A. A. 1980. Channel and hillslope geomorphology of badlands in the San Juan Basin, New Mexico. Thesis, 158 pp. Albuquerque, NM: University of New Mexico.

Hogg, S. E. 1978. The near surface hydrology of the Stevenville Badlands, Alberta. Thesis, 163 pp. Edmonton, Alberta, Canada: University of Alberta.

Jercinovic, D. E. 1984. Geomorphic analyses of small watersheds affected by coal-surface mining in northwestern New Mexico. Thesis, 244 pp. Albuquerque, NM: University of New Mexico.

Langbein, W. B., and S. A. Schumm. 1958. Yield of sediment in relation to mean annual precipitation. American Geophysical Union Transactions 39: 1076-1084.

Malde, H. E., and A. G. Scott. 1977. Observations of contemporary arroyo cutting near Santa Fe, New Mexico, U.S.A. Earth Surface Processes 2: 39-54.

Mills, A. M., and T. W. Gardner. 1983. Effects of aquifer dewatering on an ephemeral stream. *In* S. G. Wells, D. W. Love, T. W. Gardner, eds., Chaco Canyon country: A field guide to the geomorphology, Quaternary geology, paleoecology, and environmental geology of northwestern New Mexico. American Geomorphological Field Group, 57-66.

Patton, P. C., and S. A. Schumm. 1975. Gully erosion, northern Colorado, a threshold phenomenon. Geology 3: 88-90.

Schultz, J. D. 1980. Geomorphology, sedimentology, and Quaternary history of the eolian deposits, west-central San Juan Basin, northwest New Mexico. Thesis, 257 pp. Albuquerque, NM: University of New Mexico.

Schumm, J. D. 1977. The fluvial system. New York, NY: John Wiley and Sons.

Stiller, D. M., G. L. Zimpfer, and M. Bishop. 1980. Applications of geomorphic principles to surface mine reclamation in the semiarid West. Journal of Soil and Water Conservation 35: 274-277.

Thornes, J. B. 1979. Processes and interrelationships, rates, and changes. *In* C. Embleton and J. B. Thornes, eds., Process in geomorphology, 378-387. New York, NY: John Wiley and Sons.

U.S. Department of the Interior, Bureau of Land Management. 1976. Resource and potential reclamation evaluation, Bisti West study site. Bureau of Land Management report 5-1976.

Wells, S. G. 1978. Geomorphology and surface hydrology in the strippable-coal beds of northwestern New Mexico: applied to landscape reclamation. Santa Fe, NM: Office of New Mexico Energy and Minerals Department. Unpublished report.

Wells, S. G. 1982. Geomorphology and surface hydrology in the strippable-coal belts of northwestern New Mexico. Technical Report 2-68-3311, Vol. 1 and 2. Santa Fe, NM: New Mexico Energy Research and Development Institute.

Wells, S. G., and A. A. Gutierrez. 1981. Quaternary evolution of badlands in the southwestern Colorado Plateau, U.S.A. *In* R. Bryan and A. Yoir, eds., Badland geomorphology and pipe erosion. GeoAbstracts, Ltd. In press.

Wells, S. G., and D. Rose. 1981. Applications of geomorphology to surface coal-mining reclamation, northwestern New Mexico. *In* Environmental hydrol-

ogy and geology in New Mexico. New Mexico Geological Society special
publication no. 10, 69-83.
Wells, S. G., T. F. Bullard, C. D. Condit, M. Jercinovic, D. E. Jercinovic, and R.
P. Lozinsky. 1983. Geomorphic processes in the alluvial floor of the Rio
Puerco. *In* S. G. Wells, D. W. Love, and T. W. Gardner, eds., Chaco Canyon
country: A field guide to the geomorphology, Quaternary geology, paleoecology, and environmental geology of northwestern New Mexico; American Geomorphological Field Group, 37-39.

3

Pre-mining Assessments
of Reclamation Potential

Loren D. Potter

The Need for Pre-mining Assessments
of Reclamation Potential

Surface mining is an exclusive use of the land. Other land uses, such as agriculture, grazing, recreation, and habitation, are excluded during mining and for some time thereafter. The post-mining value of the land, and the quality of opportunity for land use, depends largely upon the success of reclamation. Because of the dominating influence of a surface mine on the landscape, and because of the lasting impact of the quality of reclamation, decisions regarding the management of surface mines and reclamation have special gravity.

This chapter addresses a critical component of management decisions regarding surface mining: the prediction of the ease, expense, and quality of reclamation at a given mineral-bearing site, in other words, the prediction of the *reclamation potential* for that site. Land managers must decide where to lease minerals and allow mining based on a complex assemblage of environmental, social, and economic factors. However, reclamation potential should have special weight in decisions about where to mine because the reclaimability of land is influenced by, and is influential upon, the social, economic, and physical environment of a mine site.

Role of government in pre-mining assessments

The fundamental premise for pre-mining assessments is adherence to the National Environmental Policy Act (NEPA). In general terms, NEPA requires that the impacts of projects on federal lands be anticipated, a requirement that applies to all federal leases of coal and other minable resources. Leasing decisions should consider reclamation potential on a broad geographic scale.

The requirements pertaining to individual surface coal mines are in the Surface Mining Control and Reclamation Act (SMCRA) of 1977, which is the basis for federal and state mining regulations. Both NEPA and SMCRA require that adverse impacts be *mitigated,* that is, that the severity of the impact on the environment be minimized as much as possible. Effective reclamation is the most important way to mitigate the long-term impact of a mine upon the environment.

Specifically, SMCRA requires that mine operators restore disturbed areas to a condition which is capable of supporting an equal or better use of the land than was supported prior to mining. This land use refers to the intended practical value of land for agriculture, pasturage, forestry, and so forth, which is specified in each mine's application for a mining permit. The capability and performance of reclaimed land is generally assessed in comparisons with adjacent unmined lands, or with the land as it was before mining. (For a detailed discussion of the numerical assessment of the performance of reclamation, see Chapter 9).

In much of the western United States, the principal land use is native range for grazing by livestock. Thus the primary goal of reclamation is

> to establish a productive and protective cover of durable plants, consisting predominantly of needed species adapted to and characteristic of these areas or other similar areas before mining. (Packer et al. 1979)

In this case, the success of revegetation is judged according to traditional range criteria: the abundance, vigor, and species composition of native vegetation. Accordingly, assessments of reclamation potential in the West should focus upon the ease and expense with which disturbed land can be returned to its original level of range productivity.

SMCRA requires that the land use be specified, and that the mining company present substantiating evidence that reclamation will restore the land to an equal or better condition than before mining. The mine plan must include the following ingredients:

1. Specific reclamation goals
2. Identification of problems or obstacles to those goals
3. Requirements for data to define the problems
4. The analysis of data to develop solutions
5. The procedural steps in each solution
6. The techniques for monitoring the progress of reclamation
7. The standards for measuring the success of reclamation

The reclamation program in the mine plan should be as cost-effective as possible, and should be closely integrated with the overall mining operation. Certain reclamation strategies (for example, topdressing with select overburden strata) may require that the mineral-extraction plan be adjusted and that manpower and heavy equipment be shared. The specific requirements for mine plans, and guidelines for satisfying all SMCRA regulations, are presented in a valuable handbook by Sendlein et al. (1983).

Site-specific assessments of reclamation potential should strive to anticipate as many details as possible for the reclamation program, because unexpected difficulties and expenses may reduce the cost-effectiveness, hence desirability, of mining as a land use.

Pre-mining assessments must also consider cultural resources. Remote sensing is one tool whereby the climate, vegetation, and geology of the San Juan Basin has been studied with respect to cultural resource management (Drager and Lyons 1983). Landsat images were used to construct maps of drainage patterns, geologic exposures, precipitation zones, vegetation types, human communities, and archaeological resources such as prehistoric roads and habitations. Although this chapter does not elaborate on cultural resource assessment, mineral-management decisions must consider the cultural environment as well as the physical and biological environment.

The essential problem in assessing reclamation potential

Reclamation potential may be assessed at any level on a continuous scale of specificity. Assessments for land management decisions, such as federal coal sales, should be on a broad geographic scale, involving gross characterizations and comparisons of reclaimability. Site-specific decisions regarding the licensing of a particular mine should involve a much more specific forecast of possible impediments to successful reclamation and possible sources of additional expense. In any case, the essential problem remains the same: how to study land which has not yet been disturbed in order to understand what environmental factors will dictate the reclaimability of the land once it has been disturbed. This chapter presents several approaches to this problem, including numerous examples of each.

Basic requirements for a pre-mining assessment

To judge the reclamation potential for an area, one must first understand the problems of stabilizing contoured spoils. This requires knowledge of the surface hydrology, including the erosional characteristics, of the geologic strata or soil that will be on the surface after grading. Stability depends not only on the geologic material itself, but on climatic factors such as the distribution and intensity of precipitation and wind, which may affect the timing of reclamation procedures.

Surface materials should be selected not only for their stability, but also for their capacity to support vegetation. They must provide a good seedbed, or must be improvable by mulching, fertilization, or mechanical treatment. The surface materials must also provide a favorable substrate for a diverse and durable plant community, which is reasonably resistant to natural environmental stresses such as drought or flooding.

Theoretically, a pre-mining assessment of reclamation potential must consider all those attributes of the physical environment which may influence the success of stabilization and revegetation at a given site. Table 3.1 lists climatic parameters, geologic features, and soil and spoil properties that are possible factors of importance to reclamation. Furthermore, the pre-mining assessment must consider conditions of these factors *after* mining, which may drastically alter the physical and chemical properties of the soils and geologic strata, as well as the topography and stratigraphy, at a mine site.

The best process of pre-mining assessment is that which produces the soundest management recommendations, including the most effective reclamation program. This chapter reviews several aspects of the traditional method for pre-mining assessment: the *inductive* approach, wherein baseline data (for example, Table 3.1) are collected and analyzed to induce hypotheses or predictions regarding reclaimability. Then, the reverse method is suggested: the *deductive* approach, wherein the native vegetation and the natural successional patterns are observed to deduce the influences and interactions of climate and substrate on vegetation.

For instance, when the goal of reclamation is to provide grazing habitat comparable to native rangeland, the natural vegetation can then serve as the best measure for predictions of reclamation potential and as the best guide for subsequent evaluations of reclamation success. The principal need in this case is to propose techniques which simulate the natural processes of weathering, soil formation, and vegetational development. Furthermore, these processes must be accelerated. If the time scale were not important, disturbed areas could be abandoned and left to become vegetated and eventually stabilized by natural processes.

Table 3.1. Tabulation of baseline data used to provide the basis for minimizing environmental impacts from surface mining and implementation of reclamation programs.

Climate:	*Soils and Spoils:*
Temperature	Classification
Frost-free period	Profile or depth analyses
Precipitation	Range sites
Sunlight	Texture
Wind	Water-holding capacity
Evapo-transpiration	Infiltration rate
	Stone content
(Above expressed as yearly, seasonal,	Bulk density
monthly; spatial and temporal varia-	Organic matter
bility; means; extremes; intensities)	pH
	Salt-related properties
Geology:	EC (electrical conductivity)
Regional	SAR (sodium adsorption ratio)
Physiography	Soluble cations
Elevation	Soluble anions
Drainage	Heavy metals (e.g., Se, Hg)
Hydrology	Plant macronutrients (N,P,K)
Weathering	Plant micronutrients (e.g., Zn, Cu)
Strata, chemical and physical analysis	

The Inductive Approach to Predicting Reclamation Potential

The traditional approach to assessing reclamation potential has been inductive (if reclamation science can be regarded as old enough to have any traditions). Volumes of detailed analyses of physical and biological factors have been accumulated with the rationale that the greater the amount of baseline data, the more accurate will be predictions of reclamation potential. This logic has led to substantial investments of time and money in field surveys and laboratory analyses.

Many inductive studies have produced valuable site-specific information, including management recommendations for future mining and reclamation. However, the utility of the inductive method is seriously limited by uncertainty as to which environmental factors are truly important, that is, which factors will influence the ease, expense, and quality of reclamation. One result of this uncertainty is that the data and findings from the numerous inductive studies have never been combined into a practical geographic network of information. Such a network would enable comparisons of reclamation potential across a geographic area, leading to better management decisions regarding the leasing of minerals and the licensing of mines.

Examples of inductive studies in the San Juan Basin

There have been a number of projects in coal-bearing parts of the San Juan Basin where properties of geologic strata and soils have been analyzed and interpreted to assess reclamation potential and to recommend future reclamation practices.

Overburden analyses. Rai, Wierenga, and Gould (1974) analyzed the chemical and physical properties of core samples obtained by drillings in a coal-bearing formation in the San Juan Basin. They sought to identify which geologic strata would support the best plant growth after excavation and subsequent weathering. Material from the augered cores of less than 2 mm diameter was analyzed for most of the properties in Table 3.1. The authors judged sandstones to be preferable because salinity and sodicity were relatively low and hydraulic conductivity was relatively high. Most shale layers were undesirable due to the presence of a large amount of smectite, a swelling clay. This expandable clay damages plant roots and limits infiltration of water because the surface layers, when dampened, form a water-repellent seal. Furthermore, these smectite clays are sodium-dominated, which exacerbates their water repellency and which can be toxic to vegetation.

The observations by Rai, Wierenga, and Gould (1974) were largely confirmed by a separate study at Mine B (Scholl 1982), where spoils derived from shale and sandstone had similar attributes to their shale and sandstone counterparts from the drill cores.

Regolith and soil analyses. Rai, Wierenga, and Gould (1975) also analyzed the upper layers of bedrock (regolith) and the soil horizons at coal-bearing locations. These materials were assessed for their favorability to plant growth in the context of local physiography, elevation, drainage, and certain climatic features, including seasonal patterns in temperature, precipitation, and evaporation. Soil horizons were sampled for bulk density and for field moisture content in the spring. All laboratory analyses of properties of soil and regolith (for example, Table 3.1) were on particles with less than 2 mm diameter. The authors identified the following properties to be most important in determining the suitability of these soils for plant growth: (1) availability of plant nutrients, (2) toxic concentrations of elements, and (3) physical properties affecting the movement and retention of air and water.

Severson (1981) recognized that reclamation potential in the arid Southwest may depend less on the relatively thin "A" and "B" horizons than on the "C" horizon. The upper horizons may be absent or poorly developed, and thus will constitute only a minor portion of the material which is salvaged, stockpiled, and redistributed as topdressing after mining. Severson evaluated fifty variables of the "C" horizon at seventy-seven locations across the San Juan Basin to determine how well C-horizon properties correlate with the traditional soil taxonomic classification of the overlying horizons (that is, the names of the soils according to the "seventh approximation"; Brady 1974). A strong correspondence would enable predictions of reclamation potential based strictly on the documented soil taxonomy. However, when Severson grouped the samples into different classes of suitability for use in revegetation, he observed that the classes did not correspond to the established soil taxonomy. This suggests that the taxonomic classifications of soils in the San Juan Basin may be misleading as specific indicators of reclamation potential. Severson generalized that Haplargids and Torriorthents are better than Calciorthids and Natargids for topsoiling because of fewer, if any, undesirable properties.

Problem of variability   One of the most difficult aspects in the prediction of reclamation potential is the variability within an area. If the potential is known for a soil series at one site, does that potential apply to other sites with the same series? And if there are site-to-site variations, which favorable or unfavorable properties are most variable? Those which are not important as limiting factors to reclamation may be disregarded; those which may be important deserve special attention.

Severson and Gough (1981) surveyed the geochemical variability of soils in those parts of the Basin likely to be mined. Their goals were (1) to evaluate soil variability over a broad region, (2) to appraise the chemical variability of the three soil associations most likely to be used for topdressing, and (3) to evaluate the variability in spoil and topsoil at a reclaimed surface at Mine A. The study compared the variability among cells at three different geographic scales. The dimensions of the cells were 50 by 50 km, 25 by 25 km, and 1 by 5 km. Variance ratios were calculated for different-sized cells to indicate the degree of detail required for meaningful mapping. Severson and Gough (1981) found that

> topsoil tends to be less variable in its total-element composition than does mine spoil; however, the opposite is true for extractable elements.

Total-element content reflects the type of surficial material, whereas extractable-element content reflects the pedogenic processes which are influenced by climate, topogaphy, and vegetation. More parameters were significantly variable in the C-horizon than in the A-horizon, strengthening the argument by Severson (1981) about the relative importance of the C-horizon in predictions of reclamation potential. Again, maps of the variability in element concentrations did not correspond well to maps of geologic units or soil taxonomy. Severson and Gough (1981) concluded that the variation in elemental concentrations in the Basin requires sampling at less than 5-m intervals to describe more than 50% of the total variation, which is a prohibitive level of sampling, mapping, and expense.

Transitions in properties between overburden and soil. The physical and chemical properties of a geologic strata (overburden) change as the material is excavated, shattered, and used as a soil to support vegetation. For example, pH in soils is dependent upon depth and horizon, which is not true for bedrock or spoils. In the San Juan Basin, such transitions were studied for many of the soil and spoil properties in Table 3.1 (Gould, Rai, and Wierenga 1975, Gould, Miyamoto, and Rai 1977). They found that low hydraulic conductivity, high sodium adsorption ratios, and high total salt concentrations are undesirable characteristics that do not change when an overburden is excavated and shattered. Another undesirable feature is pyrite, which is converted to sulfuric acid when exposed to air and water. The acid solubilizes aluminum, iron, and other heavy metals which become toxic to plants (Peterson and Nielson, 1973).

Gould, Miyamoto, and Rai (1977), and Yamamoto (1975) suggested that sandstone overburdens may be a reasonable alternative to natural soils due to favorable water-related properties, although the concentrations of some nutrients would be lower than in spoils derived from shales. Any nutrient deficiencies could be corrected by fertilization.

Indexing soils for reclamation potential. In New Mexico, topsoil for reclamation is defined as

> the A soil horizon layer of the three major soil horizons or other surface soil material of suitable texture and pH, and lacking concentrations of elements toxic to plants. (Mining and Minerals Division 1980)

The A-horizon of arid soils is usually thin or absent, and lacking in organic matter; but it may have a rich soil microbiota (see Chapter 8) and a granular structure which enhances aeration and penetration by roots and water. Previously, the suitability of soil for use as topdressing at New Mexico surface mines has been designated as "good," "fair," or "poor," depending on the values for several physical and chemical properties.

Daugherty (1982) analyzed properties of soils used for reclamation at Mine B to develop an index with increased accuracy for the prediction of the suitability of a soil or spoil for use as topdressing. Daugherty applied a formula for each of several soil depths, wherein a value of zero would indicate no limitations to revegetation, and an increasing positive value would indicate decreasing suitability due to one or more specific limitations. These limitations were based on the values for soil parameters known to be important to vegetation. Daugherty's index can apply to (1) individual limitations, (2) combined limitations for each depth, and (3) combined limitations for all depths. One source of uncertainty in the index, however, is the ambivalence of some soil properties. For instance, coarse fragments in soils may impair the preparation of a seedbed and thereby restrict germination, but rocks may benefit long-term production by harvesting (concentrating) water into the fine soil fraction.

Another soil index for reclamation potential is advocated by the Soil Conservation Service (1978b), which includes an economic factor. The index is based on the performance (productivity) of a local soil, less the cost of correction of any soil limitations on both a long- and short-term basis. This approach bridges the gap between predictions of the best reclamation which is theoretically achievable versus that which is technically and economically practical.

Inductive analyses of vegetation. The next logical step in the inductive approach is to assess the geochemical variability of the plant species growing on native soils and mine spoils. Gough and Severson (1981) studied concentrations of elements and variabilities in native plants growing on 1) native soils at coal-bearing parts of the San Juan Basin; 2) soils likely to be used for topdressing (for example, the Sheppard, Shiprock, and Doak series); and 3) soils at sites which have been reclaimed. Elemental concentrations were analyzed in the following species—alkali sacaton *(Sporobolus airoides)*, galleta *(Hilaria jamesii)*, broom snakeweed *(Microcephalus sarothrae)*, and four-wing saltbush *(Atriplex canescens)*—to identify potential sources of toxicity or deficiency among soils and overburdens. This study and others (for example, Gough, Severson, and McNeal 1979) have provided background concentration levels for many elements in native plant species being used for reclamation in the San Juan Basin.

There is spatial variability in the concentrations of elements in plants just as in geologic strata and soils. However, no single value for the concentration of an element can serve as a threshold for the suitability of the site for revegetation. This is because plants may vary between and within species in terms of their absorbance, accumulation, and tolerance to certain elements.

Elemental concentrations are not only important in terms of their influence upon growth rates of vegetation, but also in their susceptibility to accumulation, resulting in possible toxicity to and biomagnification in herbivores. An example is selenium, which is fixed by some plants (for instance, several species of locoweed *(Astragalus)*) and accumulated to concentrations hundreds of times greater than in the soil, thus becoming a toxic hazard to grazers on the range. Gough and Severson

(1981) found that concentrations of thirty-five elements were two to five times greater in test species growing on spoils than on native soil. In four-wing saltbush, the concentration of sodium increased proportionally to the sodium content of the growth medium. No concentrations of elements in plants, soils, or spoils, were judged to be potentially toxic to either native vegetation or to grazing cattle.

Summary of the inductive approach

The preceding examples, which are variations upon the inductive approach to pre-mining assessment, provide excellent compilations and analyses of environmental factors responsible for the species composition and productivity of plant communities on native and reclaimed surfaces. Predictions of reclamation potential are based upon assessments of the environmental conditions in light of established knowledge about the ecological tolerances of plant species. However, the numerous environmental factors need to be ranked according to their influence on revegetation and stabilization, if predictions of reclamation potential are to be more consistent, confident, and accurate. Furthermore, this ranking must recognize that environmental factors do not act as single regulators of plant growth; rather, they interact in unspecified synergistic ways to dictate the germination and growth of vegetation, that is, the success of revegetation. The fundamental limitation of the inductive approach is that predictions should account for the interactive, synergistic influence of environmental factors on vegetation, but such interactions are difficult, if not impossible, to quantify.

The Deductive Approach to Predicting Reclamation Potential

The best expression of the effect of the multitude of environmental factors on plant growth is the vegetation itself. No amount of experimentation or integration of measured factors can approach the information content of the plant's own response. This bioassay, originally called the phytometer method (Clements and Goldsmith 1924) has greatest value when entire plant communities, not just single species, are used to deduce environmental influences and processes.

The essence of the deductive method is to observe the plant community in a given environment and deduce the factors which explain the observations. The deductive method applies both the autecology of individual species and entire plant communities to interpret the processes involved in weathering or transport of exposed strata. Certain plant species, called *indicator species,* are particularly sensitive to one or more

environmental conditions; hence, they can be used to identify principal limiting factors.

The credibility of the deductive method is implied by the fact that
several of the scientists in the aforementioned inductive studies expressed satisfaction that their findings were substantiated by their observations of the dominant plants growing on the material which was
analyzed. These authors recognized that the vegetation was the ultimate standard or evaluator of the total environmental conditions.

There are two basic deductive approaches: (1) to analyze the results
of reclamation in one place and extrapolate the findings to an unmined
area of similar climate and geology; and (2) to examine the native vegetation of an unmined area on a wide variety of substrates. In both cases,
it is important to identify the goals of reclamation, as reclamation success depends in part upon the eye of the beholder. Deductive analyses
must consider whether the primary goal of reclamation is to control
erosion, to reestablish rangeland, to create farmland, or to simply return
the land as closely as possible to its pre-mining condition.

Spatial considerations in deductive analysis

Figure 3.1 illustrates three axes of spatial variation over which deductive analyses can be conducted and applied. The effect of climate on
the vegetation associated with a given geologic stratum can be studied
by observing plant communities on exposures of that stratum with differing regimes of temperature, precipitation, humidity, and so forth. The
effects of the physical and chemical properties of soils derived from different strata can be studied by observing plant communities on exposures of different strata under similar climatic regimes. Slopes provide
an opportunity to observe properties of different strata in combination,
because the transported soils there contain material from several paren-
tal strata. The geomorphic tendencies of various strata (for example,
cliff-forming versus buttress-forming behavior) are also expressed on
slopes. Other spatially varying factors which must be considered by the
deductive researcher, but which are not illustrated in Fig. 3.1, are elevation and aspect (direction of the fall line of a slope).

The San Juan Basin provides an excellent opportunity to observe the
spatial variation portrayed in Fig. 3.1. The Basin has a moderate degree
of climatic variation (described in the Introduction) and it has exposures
of folded strata which result in geologic patterns and slopes similar to
those depicted in the figure.

         Loren D. Potter

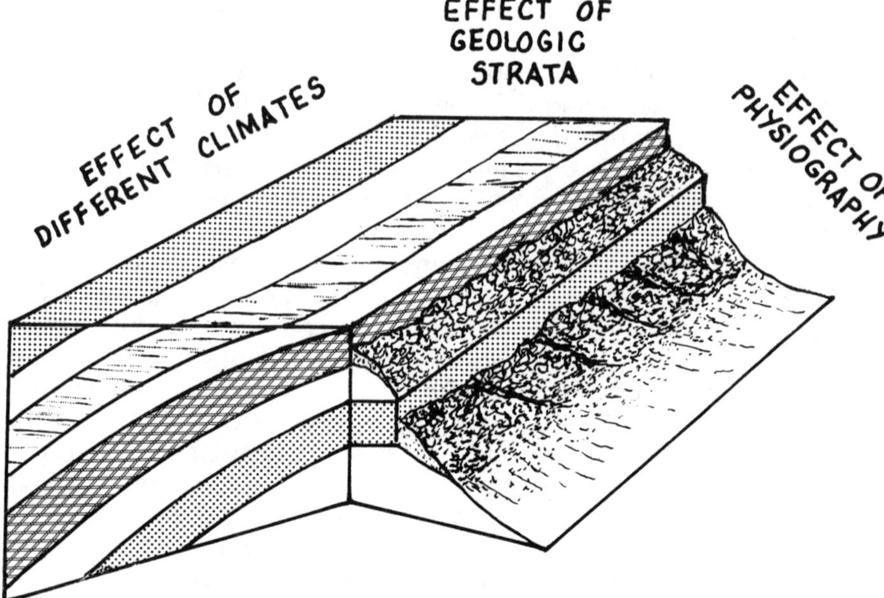

Figure 3.1. Diagrammatic representation of opportunities to study vegetational relationships to climatic variations on any given stratum; to physical and chemical variations of different strata under any given climate; and to physiography, slope, and mixing of transported soils in a region of anticlines and synclines.

Vegetation on reclaimed lands

Many studies have tried to understand reclamation and to make predictions based on analyses of post-mining stands. Unfortunately, in much of the western United States, not enough time has elapsed since revegetation practices were started to allow the vegetation to respond fully to the environment. To judge environmental influences on land reclaimed too recently would be like judging the factors responsible for a climax vegetation community by studying an early seral community.

Packer et al. (1979) attempted to predict reclamation potential by studying vegetation on lands which had been surface mined in the interior western U.S. This was a cooperative effort involving the Environmental Protection Agency, the U.S. Fish and Wildlife Service, and the U.S. Forest Service. The criteria for reclamation success were (1) the production of useful aboveground biomass (forage), which was measured with clip plots, and (2) the total stabilizing ground cover, including plant basal cover and litter, which was measured with transects. Multiple regression models were used to explain the variability among these criteria in terms of soil and spoil properties and climatic factors.

Packer et al. (1979) acknowledged the difficulty of ascribing the variation in cover and biomass to soil and climatic factors, due in part to the fact that the vegetation was not yet in balance with the environment. In their words, there was "a drastic disarray of analytical amenities." These included inconsistencies between the mines and study sites in the application rates of topsoil, seeds, fertilizer, mulch, irrigation, plus a host of other environmental and management factors. However, by using the modeling approach (see Chapter 5), the authors were able to ascribe between 50% and 70% of the variability in forage production and cover to (1) two major climatic factors (not manipulatable); (2) three properties of spoil materials (conducive to limited modification by management); (3) seven revegetation treatments, each having at least two management choices; and (4) the age of the revegetated stand.

Native vegetation before mining

The use of native vegetation as an indicator of the potential productivity of the environment overcomes the disadvantage of estimating long-term success based on the short-term results of reclamation. The native vegetation has survived and responded to stresses imposed by climatic fluctuations, physical and chemical soil properties, and grazing and trampling by wildlife and livestock.

Several studies have predicted reclamation success by studying native vegetation and soils in rangeland areas of the San Juan Basin (Gould, Howard, and Buchanan 1972, 1977; Gould, Howard and Valentine 1972). In these studies, the principal soil associations were mapped, divided into study areas, and given range site classifications. Total perennial foliar cover and species composition were derived by transects of points. The results were translated into a forage-acre factor for each site, indicating the pre-mining value and post-mining capacity of the land to support grazing. This method directly addresses the anticipated post-mine land use in predicting reclamation potential and in establishing specific standards for reclamation success.

Case study 1: Deductive approach—analyses
of the adverse properties of a marine shale

A primary goal of our reclamation research program at the University of New Mexico has been to understand the inhibitive characteristics of Mancos Shale, a geologic stratum which frequently outcrops and influences reclamation in the Colorado Plateau, the physiographic province encompassing the San Juan Basin. The program has relied principally on the deductive method, analyzing and interpreting the species composition, productivity, and physiognomy (structure) of plant communities growing on Mancos-derived soils which have been subjected to varying degrees of weathering and transportation.

Mancos shale is a Cretaceous marine shale which is stratigraphically located below the Mesa Verde Group and above the Dakota Sandstone. The shale is a common overburden in surface mining for both coal and uranium in the Southwest. It has been a problem for reclamation because of its variable pH and its high concentration of sodium, sulfate, and smectite (a swelling clay). For any given climatic area, the vegetative productivity is less on recently weathered Mancos Shale than on non-shale soils. However, there are wide variations in the nature of the vegetation communities growing on soils which are derived from the shale, but which have been variously weathered and transported (Fig. 3.2).

The objectives for the first part of the study (Potter et al. 1985a) were to sample the vegetation on different soils derived from Mancos and to relate the observations to the following factors: (1) physical and chemical properties of the soil; (2) physiographic location; (3) admixtures and influences of other materials such as alluvium and sandstone; and (4) the lithology of the parent shale.

A collection was made of 120 bedrock and 123 soil samples from Arizona, Utah, Colorado, and New Mexico. Profiles were described for shale-derived soils in a variety of topographic situations. We also observed the plant communities on the sampling sites and on adjacent soils derived from sandstone, to provide a contrasting example of the environmental capacity to support vegetation. This sampling scheme yielded data for the same soils in different climates and for different soils in the same climate, enabling analysis of the spatial patterns described in Fig. 3.1

Observations and deductions regarding variability.   The parent shales themselves showed great variety in physical characteristics, ranging from solid, blocky, fine-textured gray shale to friable shale breaking into thin plates. There was also great chemical variability in concentrations of iron oxides, white gypsum crystals, and yellow deposits of sulfur or jarosite (potassium aluminum iron sulfate). The presence of sulfur caused pH to be as low as 3.4; conversely, the presence of calcium carbonate raised pH as high as 9.4. The above variations were reflected by concomitant variation in the vegetation on shales within one climatic regime.

Interbedded layers of sandstone, which may occur anywhere in the

Figure 3.2. Caprock of Mesa Verde Group overlaying a deep layer of Mancos Shale. Note the sandstone rubble on erosional slopes and the variations of vegetation on the outwash slopes due to the varying amounts of added sand to the weathering and leaching of the transported shale.

Mancos, affect both topography and vegetation. At many locations, erosion has removed overlying shale to expose a sandstone caprock, a platform which temporarily protects the underlying shale from further erosion and which gradually disintegrates to coarse fragments and sand that alter the texture, water relations, and vegetational growth downslope. Fracture lines in the sandstone are often continued into the underlying shale, causing yellow bands of weathered shale for several inches alongside the fracture. Evening primrose *(Oenothera* sp.) grows on these weathered bands, but not on adjacent shale, indicating that weathering removes some factor which is inhibitive to the growth of this plant.

The mantle of sandstone fragments downslope from a caprock appears and acts like desert pavement, resisting erosion and improving water relations by reducing erosion and concentrating water into the fine soil fraction ("water harvesting"). The sandstone also alters the texture of soils downslope, adding sand to the silt and clay particles, thereby improving aeration and permeability to water.

Interbedded strata of bentonite in the shale have nearly the opposite effects as interbedded sandstone layers, decreasing infiltration of water and inhibiting vegetation. Bentonite, which is a clay derived from volcanic ash, is very high in smectite, which swells and prevents water from infiltrating and leaching inhibitory levels of sodium and sulfate. Thick layers of bentonite frequently form erosion-resistant surfaces barren of vegetation.

Observations and deductions of vegetation and physiography   We used transects to observe, interpret, and illustrate patterns in the vegetation on slopes at different exposures of Mancos Shale. Figure 3.3 contains two idealized transects that synthesize observations from transects in all but the most arid regions of our study area.

Figure 3.3A illustrates how ponderosa pine *(Pinus ponderosa)* occupies fractured sandstone caps and rubble on upper slopes in relatively cool and moist regions of the Colorado Plateau. The climax vegetation for these regions is pinyon *(Pinus edulis)* and juniper *(Juniperus* spp.) woodland, but the sandstone improves the moisture supply sufficiently to support a more mesophilic ponderosa pine forest with associated understory. A similar phenomenon explains why ponderosa pine occurs on rocky knolls in the semiarid grasslands of the western Great Plains, even though laboratory analyses of the soil fractions from knoll and grassland sites show no significant differences in physical and chemical properties (Potter and Green 1964, Potter 1969). The fractured rocks improve water relations by preventing the surface from becoming sealed, thereby allowing water to infiltrate more deeply, which, in turn, leaches soluble salts and encourages deeper penetration by roots.

In the same regions as dicussed above, exposed gentle slopes are often occupied by pinyon-juniper, oak brush *(Quercus* spp.), and mountain-mahogany *(Cercocarpus* spp.). These vegetation types are favored by a thin layer of shale over fractured sandstone, or by eroded sandstone rubble and talus. In both situations, the presence of sandstone improves infiltration, aeration, rooting, and water relations in general.

Figure 3.3B illustrates vegetational patterns in more arid regions in the Colorado Plateau, including the San Juan Basin. In these areas, ridges capped with sandstone support pinyon and Utah juniper *(Juniperus osteosperma)*. Junipers also occur at lower elevations where sandstone fragments have accumulated in drainages where conditions are relatively mesic, not only because of the sandstone but because of the greater weathering and leaching of the underlying shale. Alluvial gravel or outwash overlying the shale will support oak brush, big sagebrush *(Artemisia tridentata)*, New Mexico olive *(Forestiera neomexicana)*, and rabbitbrush *(Chrysothamnus* spp.). Lower in the outwash, where a mantle of sand covers the shale, four-wing saltbush and Indian ricegrass

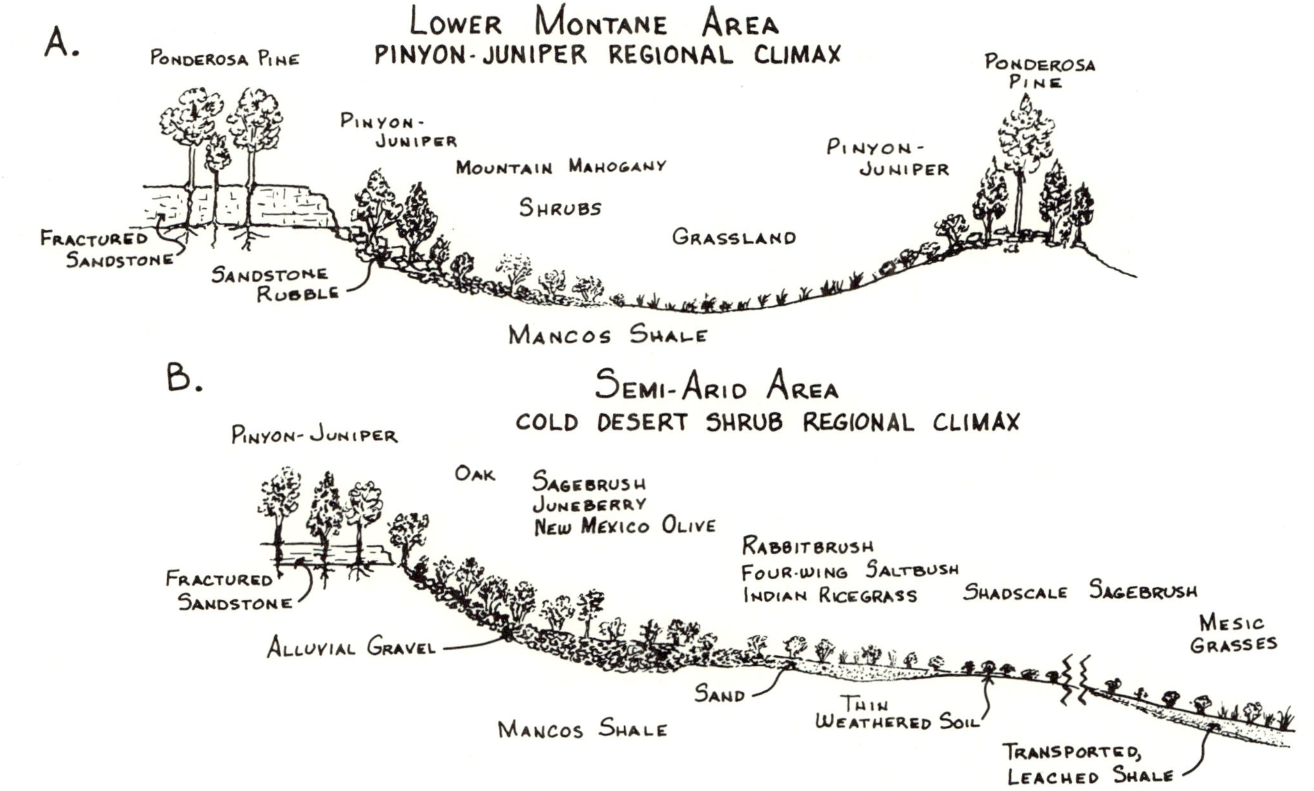

Figure 3.3. Plant communitites as related to physiography and amendment of Mancos Shale by additions of sandstone or coarse alluvium or by distance of transport. A. is in the lower montane area; B. is in the semiarid area.

(Oryzopsis hymenoides) grow; these two species, particularly the latter, are indicators of the presence and influence of sand. Where the mantle is absent, the dominant plant will be shadscale *(Atriplex confertifolia)*, a stunted shrub which grows only where sodicity and salinity inhibit most other species. However, the concentrations of sodium, sulfate, and other soluble ions may be reduced somewhat if the weathered shales are transported and leached, resulting in a deep, calcareous, well-structured, silty clay alluvium that may support big sagebrush or a rich valley grassland. By contrast, residual soils on steep shale slopes in the same climate will be virtually barren.

Interpretation of vegetation and soils along transects.   Five transects in different vegetation types and physiographic situations received a combination of vegetation sampling and laboratory soil analyses (Potter et al. 1985b). Figure 3.4 illustrates such a transect from an arid portion of the Colorado Plateau (10 cm annual precipitation) where the vegetation is dominated by species of the arid and saline Great Basin flora. The upper slope supports about 30% cover of blackbrush *(Coleogyne ramosissima)* and a 10% cover of galleta and fescue *(Festuca* sp.). Sodium and sulfate concentrations in the soil are low and the surface is friable and sandy due to the influence of fragments of Mesa Verde Sandstone. There is a sharp shift to shadscale in the vegetation downslope, which seems almost entirely due to increased sodium in the subsurface of the soil. Further downslope, horsebrush *(Tetradymia spinosa)* codominates with shadscale on thin, loose, silty clay soils. Still further downslope, mat saltbush *(Atriplex corrugata)* takes over, and subsurface sulfate increases from 30 to 1650 ppm over 4 m distance. Galleta decreases with this increase in sulfate concentration, a phenomenon observed elsewhere between paired samples from sites with and without galleta.

Where horsebrush is absent, leaving only mat saltbush, the soil is only 2 to 5 cm thick over a flaky shale bedrock. In the downslope portion of the mat saltbush habitat, soils have a biscuit-like surface which is indicative of high sodium content. Mat saltbush was found to tolerate sodium concentrations of 2,524 ppm at the surface and 17,685 ppm in a saturated solution of the subsurface soil.

Relationship of vegetation to slope and aspect.   The vegetation communities on the same type of shale may differ depending on the direc-

---

OPPOSITE PAGE:

Figure 3.4.   Relationships of vegetational changes to soil characteristics along a gentle slope of Mancos Shale where the upper slope is affected by the additions of weathered sand, in an arid region typical of the San Juan Basin.

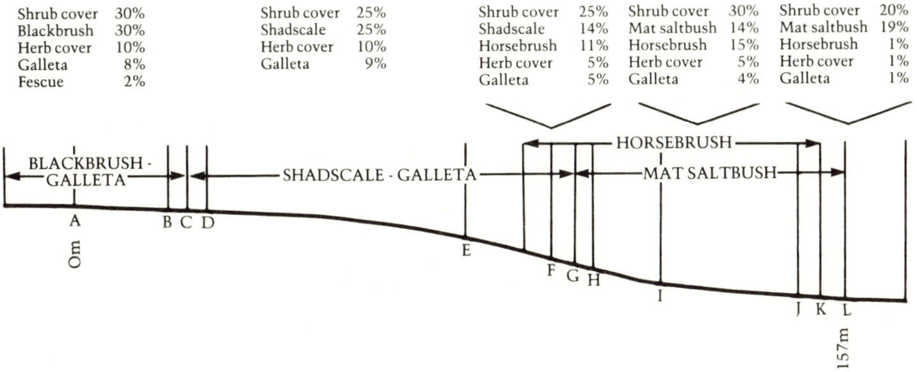

	A		B		C		D	
Depth (in)	0-2	8	0-2	4	0-2	6	0-2	6
Sand/clay(%)	56/16	—	—	60/23	59-14	—	—	39/36
pH	8.3	8.5	8.8	8.9	8.8	9.1	8.9	9.4
Cond.(mmhos/cm)	.2	.1	.4	.2	.2	.2	.2	.2
Carbon (%)	.71	.45	.72	.52	6.9	.44	.50	.54
Sulfate (ppm) (1:5 extract)	<30	—	<30	<30	<30	<30	—	—
G.S.I.	—	—	—	—	—	—	—	—
Sodium (ppm) (sat. soil sol,n.)	13	27	27	13	13	202	27	229

	E		F		G		H	
Depth (in)	0-2	6-8	0-2	6-8	0-3	7	0-2	5
Sand/clay (%)	62/12	55/24	—	9/52	7/50	10/51	28/39	—
pH	8.9	9.1	9.1	9.1	9.0	7.8	9.1	8.0
Cond.(mmhos/cm)	.3	.2	.2	.1	.1	2.5	.3	2.3
Carbon (%)	.44	.46	.42	.25	.36	.36	.89	.35
Sulfate (ppm) (1:5 extract)	<30	<30	<30	<30	—	1650	~40	1400
G.S.I.	—	—	—	—	—	1.21	—	0.98
Sodium (ppm) (sat. soils sol,n.)	27	27	27	54	27	81	54	81

	I		J		K		L	
Depth (in)	0-2	6	0-2	4	0-2	3-4	0-2	5
Sand/clay (%)	11/41	—	12/49	4/85	—	—	9/49	4/9
pH	8.0	7.8	8.9	8.2	8.2	8.1	8.1	8.4
Cond.(mmhos/cm)	2.3	2.3	.3	2.7	2.3	2.2	2.4	3.1
Carbon (%)	.37	.30	.44	.53	.44	.39	.37	.41
Sulfate (ppm) (1:5 extract)	1300	1300	~40	2080	1200	1750	1520	2900
G.S.I.	0.89	1.07	—	1.59[+]	0.81	1.29	1.17	2.37[+]
Sodium (ppm) (sat. soils sol,n.)	13	27	67	351	13	108	121	553

tion of the slope exposure. For instance, southwest-facing slopes may support shadscale and mat saltbush while northeast-facing slopes support big sagebrush. Reclamation managers should anticipate that the post-mining topography may cause variability in the germination and productivity of a uniform planting program. One option in practice at Mine C in the San Juan Basin is to seed southerly slopes with species that are adapted to relatively stressful moisture conditions on such exposures.

In relatively level areas without sandstone or bentonite, residual soils develop from the weathering of Mancos Shale *in situ*. Pedogenesis does not change the clay mineralogy, which is about 50% sodium-dominated smectite in Mancos. The limiting factors for such a soil in arid or semi-arid conditions are poor infiltration, drought, sodicity, and salinity, as evidenced by the fact that only drought- and salt-resistant species grow on residual Mancos soils. These or similar soils, if encountered in reclamation, should be amended to reduce the inhibitory effects of excess clay and soluble salts.

Identification of inhibitive factors. To clarify and confirm some of our deductive observations about Mancos Shale in the field, Louderbough (1983) conducted growth studies in the laboratory and greenhouse. The specific goal of the study was to resolve the influences of physical properties, principally texture, versus chemical properties, principally the osmotic content of the soil solution.

Western wheatgrass *(Agropyron smithii)*, a common range grass in the San Juan Basin, was grown in hydroponic solutions made from four different shale samples, thus eliminating the effect of texture and isolating the chemical variability among the shales. For each kind of shale, the grass was also grown in soils with textures that had been artificially modified by differential additions of washed silica sand. All three of the resulting textural classes (silty clay, sandy clay, and sandy loam) received the same amount of water, which was that required to bring the sandy loam to field capacity twice a week.

The sandiest texture produced the best growth for each kind of shale, substantiating our field observations about the favorable effects of sandstone. The shale with the highest concentrations of cations and sulfates produced the best growth when modified with sand. This led to the conclusion that texture outweighs chemistry as a determinant of the vegetation, and the revegetation potential, of Mancos Shale.

Management recommendations based on deductions. The most important recommendation is to modify the textures of Mancos soils or overburdens by adding calcareous sand or coarse fragments to improve water relations. These measures will improve aeration, infiltration, and drainage, which will in turn ameliorate chemical adversity by leaching soluble salts. Furthermore, the addition of sand or coarse fragments will be much easier and less expensive than complex chemical manipulations.

The leaching of adverse chemicals from sodic and saline spoils may also be accomplished by adding organic matter (for example, crimping in straw) to improve structure, or by irrigating (May 1975, Merino and Crookston 1977, Thames and Verma 1977, Aldon 1978). Other measures include the addition of calcium sulfate or sulfuric acid to replace sodium with calcium on the adsorption sites, thereby reducing the dispersion of colloids and increasing soil structure (Kelley 1964, Fletcher and Schultz 1975, Ryan et al. 1975). These measures essentially mimic or accelerate the natural weathering process of Mancos Shale.

Case study 2: Deductive approach—selection and manipulation of a favorable stratum

A second goal of our reclamation research program at the University of New Mexico was to use the deductive approach to select and manipulate an overburden stratum so as to achieve the most favorable reclamation possible at a limited cost. The work was done at Mine E, an open-pit uranium mine in the southeast corner of the San Juan Basin, where drought and poor soils inhibited reclamation.

The research, which was conducted by Kelley (1979), began with observations of the adjacent native range and the invasion by weeds onto banks of raw spoil. The environment of the mine consisted of mesas, plateaus, and steep slopes with thin, rocky soils supporting blue grama *(Bouteloua gracilis)* and galleta grassland with scattered one-seed juniper *(Juniperus monosperma)* and pinyon on the rocky, north-facing slopes. Deeper, finer alluvial soils support alkali sacaton grassland with scattered four-wing saltbush, shadscale, or greasewood *(Sarcobatus vermiculatus)*.

Kelley observed that the most mesic native species grew on fragmented, rocky exposures with natural water-harvesting characteristics. The greater the percentage of soil occupied by rock, the more the natural water supply would be concentrated into the remaining fine soil fraction. The increased concentration of water, plus the fact the surficial rocks impede evaporation, greatly benefits the productivity of vegetation in arid environments.

Field observations and lab analyses. The overburden at Mine E was Mancos Shale with interbedded layers of sandstone called Tres Hermanos Sandstone. Kelley's observations of spoil dumps, some of which had not been disturbed for twenty years, indicated that natural plant invasion was greatest and natural succession had proceeded farthest on sandstone rubble. Soil analyses (for example, Table 3.1) which were conducted on material from the dumps indicated that Tres Hermanos Sandstone clearly had the best properties for the establishment of native vegetation and the restoration of range. Also, Tres Hermanos promised to be easy to crush into smaller fragments.

Preparation of seedbed.   The decision was made to spread the sandstone boulders and rocks over the dump surface and to crush them into smaller fragments with bulldozers. This was judged equal or better than the very costly alternative of salvaging the scarce alluvial soils from adjacent valleys, which would themselves require reclamation after the removal of soil. Most rock was crushed to a coarse soil-like matrix, although large boulders were pushed aside to form erosion-resistant rip-rap berms or clumped to produce rocky mounds for wildlife. Then, the surface was treated with a 44-ton compactor, further crushing the surface particles to a size which could be seeded with a range drill. The compactor tracks pocked the surface, leaving small catchment basins to collect rain and aeolian material, as well as to resist erosion. The final material contained 20% to 40% rocks. The cost of the whole process was judged to be one-third or less the cost of salvaging adjacent alluvial soil (Reynolds et al. 1978).

Revegetation of crushed overburden.   The area was seeded with a mixture of nine locally abundant grasses and one legume species. Barley straw was crimped into the surface. The barley mulch provided an unexpected benefit by germinating in the fall, dying back, regrowing from the rootstock like a winter annual, and producing seed. The barley did not persist in the community, but it provided a valuable "early successional" input of organic matter into the coarse soil. The surface was not irrigated, even though the second growing season had only 65% of the usual precipitation.

The density and coverage of vegetation was monitored in quadrats along permanent transects. The most successful grass was side-oats grama *(Bouteloua curtipendula)*, a mesic species for the area. Alkali sacaton increased in abundance over time. Within two years, four-wing saltbush and other browse shrubs invaded the plots from surrounding areas and produced seed. The species composition and productivity of the two-year-old plant community on the reclaimed surface compared favorably with nearby native communities. The experimental reclamation program seemed to fulfil our principal ambition, which was to accelerate the process of primary natural succession, including soil formation, on a severely disturbed surface.

Improving Pre-mining Assessments of Reclamation Potential with the Deductive Approach

So far, the government has relied principally upon inductive analyses to predict reclamation potential in the San Juan Basin. The preparation of management documents, such as Environmental Impact Statements, has been preceded by extensive and expensive environmental surveys.

For instance, the San Juan Basin has been the site of numerous EMRIA (Energy Mineral Rehabilitation Inventory and Analysis) surveys, each of which has generated a report exceeding 250 pages in length (for example, the 1976 U.S. Department of Interior report for the Bisti West Study Site). This program was a joint venture by the Bureau of Land Management, the Bureau of Reclamation, and the U.S. Geological Survey, and was mandated to assemble adequate baseline data for the selection of reclamation goals and the establishment of lease stipulations. The program amassed detailed information on meteorology and climatology; physiography and drainage; regional and site-specific geology; coal resources; soil and bedrock properties; land uses; vegetation types and range classifications; plant-soil water relationships; infiltration and soil detachability; and hydrology, including the supply of water for irrigation. All of the information in the EMRIA program, along with data and findings from a host of other government-sponsored projects, were to be used to increase the accuracy and effectiveness of Environmental Analyses (EA), Environmental Impact Statements (EIS), and the entire program for exploration, planning, and development of coal in the San Juan Basin.

In retrospect, it turns out that little, if any, of this enormous database was actually used in the impact assessment and planning process for the Basin. The failure to use this expensive and potentially informative assemblage of environmental data, and the intrinsic liability of the inductive method for pre-mining assessment, can be understood by examining the steps and rationales for the technique.

Liabilities of the inductive approach

Generally speaking, the inductive approach contains the following steps: (1) the observation and documentation of all facts, (2) the analysis and classification of these facts, (3) an inductive derivation of generalizations from the facts, and (4) further testing of the generalizations.

Traditionally, steps 1 and 2 are not accompanied by hypotheses, which results in considerable effort being wasted on the collection of irrelevant data and the analysis of irrelevant facts. However, the fundamental fallacy of the technique is that all the relevant facts can never be observed without prior formulation of hypotheses, that is, there is no way to collect all the relevant data without a knowledge of the hypotheses to which the data have relevance.

Another weakness is that no general rules describe how to induce hypotheses from empirical data. Progress from data to hypotheses requires creative invention, not systematic derivation, and this causes inconsistencies among scientists. Once formulated, the hypotheses must be validated. Such validation is generally achieved by the same general procedures as used for confirmation of hypotheses which have been formulated by deduction.

The advantage of the inductive procedure is that data is gathered for numerous parameters in a variety of ecosystems, which adds to scientific knowledge of the environment. However, for management purposes, pre-mining assessments have so far been little more than aimless accumulations of environmental data. Then, as often happens, analysts of environmental impacts either fail to cope with the intimidating volume and mixture of data or they, somewhat ironically, complain that more data are needed, although failing to specify guidelines for the collection of additional data.

Advantages of the deductive approach

The deductive approach promises to streamline the collection of baseline data required for sound management. Hypotheses and predictions of reclamation potential, which would be based on deductive observations of natural vegetation and weathering patterns, could be verified with relatively few, selected site-specific environmental inventories and analyses. Deductive hypotheses have the further advantage that the conclusions are related to the premises in such a way that, if the premises are true, the conclusion must also be true. This gives direction to the investigation.

Limitations of the deductive approach

The principal limitation of the deductive method is its need for a scientist or land manager who is sufficiently experienced and observant to detect and interpret natural patterns of vegetation and geologic weathering. The method also requires a certain degree of spatial variation in geology and climate, as illustrated in Fig. 3.1; the method would not be applicable to flat landscapes with horizontal strata.

Summary

Most pre-mining assessments never grasp the real issue. They anticipate neither the real ecological impacts of the mining operation, nor the nature of the reclamation recommendations which will be required.

An effective deductive pre-mining assessment must anticipate the above factors, as well as the specific goals of reclamation, including the post-mining land use.

If baseline data intended for inductive analyses have already been gathered, then relevant information can be analyzed to deduce that which is necessary for predictions of reclamation potential. Alternative predictions can then be validated by inference or experimentation.

In future efforts, time and money can be saved by observing the most productive vegetation in a given climate or on a given geologic substrate, and then deducing the factors of importance to reclamation. This productive vegetation can provide a basis for predictions and a standard for reclamation success. This procedure will save time and expense, as well as allowing predictions of reclamation success in areas which have been neither mined nor reclaimed.

References

Aldon, E. F. 1978. Reclamation of coal-mined land in the Southwest. Journal of Soil and Water Conservation (March-April):75-79.

Brady, N. C. 1974. The nature and properties of soils. New York, NY; MacMillan Publishing Co.

Clements, F. E., and G. W. Goldsmith. 1924. The phytometer method in ecology. Washington, DC: Carnegie Institute Publication 356.

Daugherty, L. 1982. Selecting topsoil for coal mine reclamation. *In* E. F. Aldon and W. R. Oaks, eds., Reclamation of mined lands in the Southwest, 50-56. Albuquerque, NM: New Mexico Chapter, Soil Conservation Society of America.

Drager, D. L., and T. R. Lyons. 1983. Remote sensing in cultural resource management: The San Juan Basin Project. Washington, DC: Cultural Resource Management Division, National Park Service.

Fletcher, J. R., and G. D. Schultz. 1975. Sulfuric acid as a soil amendment to enhance plant growth. Mining Congress Journal (August):34-39.

Gough, L. P., R. C. Severson and J. M. McNeal. 1979. Extractable and total-soil element concentrations favorable for native plant growth in the northern Great Plains. *In* M.K. Wali, ed., Ecology and coal resources development, 859-869. New York, NY: Pergamon Press.

Gough, L. P., and R. C. Severson. 1981. Biogeochemical variability of plants at native and altered sites, San Juan Basin, New Mexico. U.S. Geological Survey professional paper 1134-D.

Gould, W. L., V. W. Howard, Jr., and B. A. Buchanan. 1972. Biotic composition and soil characteristics of an area to be strip-mined in San Juan County by El Paso Natural Gas Company. Las Cruces, NM: New Mexico State University Agricultural Experiment Station special report no. 24.

Gould, W. L., V. W. Howard, Jr., and B. A. Buchanan. 1977. Biotic composition and soil properties on a Fruitland Formation exposure in northwestern New

Mexico. Las Cruces, NM: New Mexico State University Agricultural Experiment Station special report no. 33.

Gould, W. L., V. W. Howard, Jr., and K. A. Valentine. 1972. Soil characteristics, biotic composition and vegetative production of areas leased by Western Coal Company for strip mining near Fruitland, New Mexico. Las Cruces, NM: New Mexico State University Agricultural Experiment Station special report no. 20.

Gould, W. L., S. Miyamoto, and D. Rai. 1977. Reclamation potential of overburden materials in the Fruitland Formation in the San Juan Basin. Las Cruces NM: New Mexico State University Agricultural Experiment Station bulletin no. 657.

Gould, W. L., D. Rai, and P. J. Wierenga. 1975. Problems in reclamation of coal mine spoils in New Mexico. *In* M.K. Wali, ed., Practices and problems of land reclamation in western North America, 107-121. Grand Forks, ND: The University of North Dakota Press.

Kelley, W. P. 1964. Review of investigations on cation exchange and semiarid soils. Soil Science 97:80-88.

Kelley, N. E. 1979. Vegetational stabilization of uranium spoil areas, Grants, New Mexico. Los Alamos, NM: Los Alamos Scientific Laboratory publication LA-7624-T.

Louderbough, E. T. 1983. Chemical and physical characteristics of Mancos Shale soils and their inhibitive effects on plant growth. Dissertation, 86 pp. Albuquerque, NM: University of New Mexico.

May, M. 1975. Moisture relationships and treatments in revegetating strip mines in the arid West. Journal of Range Management 28:334-335.

Merino, J. M., and R. B. Crookston. 1977. Reclamation of spent oil shale. Mining Congress Journal 63(10):31-36.

Mining and Minerals Division. 1980. State of New Mexico surface coal mining regulations: Rule 80-1. Santa Fe, NM: The Energy and Minerals Department, State of New Mexico.

Packer, P. E., C. E. Jensen, E. L. Noble, and J. A. Marshall. 1979. Estimating revegetation potentials of land surface mined for coal in the West. *In* M.K. Wali, ed., Ecology and coal resource development, 396-411. New York, NY: Pergamon Press.

Peterson, R. F., and H. B. Nielsen. 1973. Establishing vegetation on mine tailings waste. *In* Ecology and reclamation of devastated land, 3-14. New York, NY: Gordon and Breach.

Potter, L. D. 1969. Ponderosa pine-grassland competition. *In* C.C. Hoff and M.L. Riedesel, eds., Physiological systems in semiarid environments, 285-286. Albuquerque, NM: University of New Mexico Press.

Potter, L. D., and D. L. Green. 1964. Ecology of ponderosa pine in western North Dakota. Ecology 45:10-23.

Potter, L. D., R. C. Reynolds, and E. T. Louderbough. 1985a. Mancos Shale and plant community relationships: I. Field observations. Journal of Arid Environments 9:137-145.

Potter, L. D., R. C. Reynolds, and E. T. Louderbough. 1985b. Mancos Shale and plant community relationships: II. Analysis of shale, soils, and vegetation transects. Journal of Arid Environments 9:147-165.

Rai, D., P. J. Wierenga, and W. L. Gould. 1974. Chemical and physical properties of core samples from a coal-bearing formation in San Juan County, New Mexico. Las Cruces, NM: New Mexico State University Agricultural Experiment Station research report no. 287.

Rai, D., P. J. Wierenga, and W. L. Gould. 1975. Chemical and physical properties of soil samples from a coal-bearing formation in San Juan County, New Mexico. Las Cruces, NM: New Mexico State University Agricultural Experiment Station research report no. 294.

Reynolds, J. F., M. J. Cevik, and N. E. Kelley. 1978. Reclamation at Anaconda's open-pit uranium mine. Reclamation Review (Canadian Land Reclamation Association), 1(1):9-17.

Ryan, J., J. L. Strochlein, and S. Miyamoto. 1975. Effect of surface-applied sulfuric acid on growth and nutrient availability of five range grasses in calcareous soils. Journal of Range Management 28:411-414.

Scholl, D. G. 1982. Properties of coal spoils of the Navajo mine in northwest New Mexico. *In* E. F. Aldon and W. R. Oaks, eds., Reclamation of mined lands in the Southwest, 35-42. Albuquerque, NM: New Mexico Chapter, Soil Conservation Society of America.

Sandlein, L. V. A., H. Yazicigil, and C. L. Carlson. 1983. Surface mining environmental monitoring and reclamation handbook. New York, NY: Elsevier.

Severson, R. C. 1981. Evaluating chemical character of soil material for suitability in rehabilitating mined land in the San Juan Basin, NM. Soil Science Society of America Journal 45(2):396-404.

Severson, R. C., and L. P. Gough. 1981. Geochemical variability of natural soils and reclaimed minesoils in the San Juan Basin, NM. U.S. Geological Survey professional paper 1134-C.

Soil Conservation Service. 1978. Soil potential ratings. National Soils Handbook notice 31. Washington, DC: U.S. Department of Agriculture, Soil Conservation Service.

Thames, J. R., and T. R. Verma. 1977. Reclamation on the Black Mesa of Arizona. Mining Congress Journal 63:42-46.

U.S. Department of Interior. 1976. Resource and potential reclamation evaluation: Bisti West Study Site. Washington, DC: EMRIA report no. 5.

Yamamoto, T. 1975. Coal mine spoil as a growing medium: Amax Belle Ayr South Mine, Gillette, Wyoming. *In* Third symposium on surface mining and reclamation, Vol. 1, 49-61. NCA/BCR Coal Conference and Exposition 2, Louisville, KY. Washington, DC: National Coal Association.

4

The Role of a Government Experiment Station and the Experimental Approach in Developing and Testing Reclamation Practices

Earl F. Aldon

Government and Environmental Protection

Natural resources have been severely disturbed by man throughout our recent history. Federal legislation has responded to public concern toward minimizing the impact or reversing the consequences of such disturbances. For instance, the U.S. Department of Agriculture (USDA) has established several agencies in response to recognized public needs. As a result of public concern about the destruction of eastern forests through excessive timber harvesting and burning, the Forest Service was created in 1905 and charged with the administration of the National Forest Preserves. The Soil Conservation Service was created because of the dust bowl conditions in the Great Plains in the 1930s. More recently, the Environmental Protection Agency and the Office of Surface Mining (in the Department of Interior) have been established in response to specific resource problems.

This chapter describes how the Rocky Mountain Forest and Range Experiment Station, which is part of a geographic network of research facilities supported by the USDA Forest Service, has identified and responded to needs for reclamation research in the San Juan Basin. The Experiment Station's research procedures and activities are described in general terms, and then illustrated with a case study involving plant-establishment techniques for alkali sacaton *(Sporobolus airoides),* a desirable native grass for reclamation. A complete review of the principal laws which direct Forest Service activities is provided by the U.S. Department of Agriculture (1983).

The chapter also describes the experimental approach to reclamation research. Experiments are scientific procedures which identify and quantify relationships between variables in nature. In an experiment, a scientist varies the value of one factor and measures the effect, if any, on some other factor.

The Experiment Station uses the "scientific method," which is a systematic *format* for the design, execution, and interpretation of experiments. The experiments yield results with sufficient confidence to allow immediate translation of the findings into specific management recommendations.

Government Conduct of Reclamation Research

For twenty years prior to 1973, personnel at the New Mexico Office of the Rocky Mountain Forest and Range Experiment Station had been conducting research directed at rehabilitating overgrazed and depleted rangelands on the Rio Puerco drainage. In 1973, because of public concerns, our research activities were expanded to include reclamation of lands disturbed mainly as a result of surface mining. Our scientists and Assistant Director prepared a Research Work Unit (RWU) description containing the following elements: a mission statement, a justification of need, a selection of specific problems, an approach to the solution of those problems, and a proposed location for the application of the research. The RWU description was then widely distributed for comment from concerned individuals in universities, federal and state agencies, Indian tribal governments, environmental groups, and industry.

After considerable review and feedback, a final RWU description was prepared. This document was approved by the Experiment Station Director with the concurrence of other USDA officials. The RWU was a five-year charter for the conduct of specific research on the reclamation of disturbed lands in the Southwest. The Project Leader then prepared a Problem Analysis which allocated staff and costs to the various studies over the five-year time frame, and which was reviewed and approved by the Assistant Director. The final step prior to beginning the actual research was the preparation of detailed Study Plans which included the following: a literature review, a statement of objectives, an experimental design, a forecast of results, a budget including labor, and an intended scientific or technical publication outlet.

Reclamation research problems in the arid Southwest

The Albuquerque Field Unit of the Experiment Station identified four major elements in the development of methods for establishing and maintaining productive ecosystems on mined lands in the Southwest: (1) the identification of a diverse array of plants adapted to spoil environments;

(2) the development of seeding and planting methods; (3) the design of post-mining landscapes to maximize stability, limit water runoff, and enhance revegetation; and (4) the development of methods for managing reclaimed sites.

Identification of a diverse array of plants adapted to spoil environments. This element begins with studies of climatological data and of the physical and chemical properties of spoil materials. This information is combined with established knowledge of the physiology of candidate plants to assess the probable adaptability of various species. Promising shrubs and forbs, which are available from SCS Plant Material Centers (see Chapter 6), are tested for germination, early growth, and survival on various coal spoil materials. Strong candidates are directly seeded into spoils in the field. Others are "outplanted," meaning that nursery-grown stock is planted into field spoils. Finally, decisions are made on the suitability of various species based on reproduction, growth rate, colonization behavior, nitrogen-fixing abilities, and responses to management alternatives such as irrigation and surface amendments.

Development of seeding and planting methods. This element includes seedbed preparation, seeding rates, and dates and methods of seed-sowing to ensure the establishment of stands by direct seeding. Also involved are methods of producing and outplanting stock of herbaceous and woody species with a relatively high expectation of survival, reproduction, and subsequent stabilization of coal spoils. Research on methods of outplanting encompasses the physiology of the plants and the microbiology of the rhizosphere. The performance of outplant stock is evaluated according to the initial growth media, the conditions of the nursery climate, the nutrient and water regimes, the age at outplanting, the techniques of outplanting, and any "conditioning" of the outplant stock (the induction of physiological hardening in plants to prepare for the stress of transplantation).

Design of post-mining landscapes to maximize stability, limit water runoff, and enhance revegetation. This element concerns the effect of the length and steepness of slopes upon the erodibility of spoil by wind, water, and gravity. Also involved is the suitability of various spoil and overburden components as growth-support media, since vegetation stabilizes spoil. Stratification schemes (for example, sandstone-derived soils over shale overburden) are tested for their stability and ability to supply water and nutrients to vegetation. Studies include the experimental shaping of topography to trap winter moisture, for instance, by cutting furrows at regular intervals along the contours of slopes. Water movement in spoil banks is tracked, and chemical properties of runoff are monitored.

Development of methods for managing reclaimed sites. This element considers the use of rehabilitated land by livestock and wildlife, which must be adjusted according to the requirements for the establishment and maintenance of vegetation. Management plans must recognize that reclaimed areas may be more sensitive than adjacent rangeland and that special standards govern their long-term rehabilitation. The lushness and palatability of seeded and planted vegetation may attract sufficiently large numbers of animals, such as deer, rodents, and rabbits, to damage the stand. Measures such as fencing, poisons, and repellents are tested for their ability to protect areas susceptible to invasion by large numbers of wildlife.

The stepwise testing and demonstration of reclamation methods

The reclamation research program at the Rocky Mountain Forest and Range Experiment Station progresses in a stepwise fashion from small scale to large scale. At each step, a number of alternatives are compared. Only those alternatives which perform best are carried forward for further consideration. The procedure has been likened to the pruning of a tree (Platt 1964), where branches which are not promising are removed, thereby encouraging the growth of the remaining branches.

There are four general steps or levels at which the Experiment Station tests and demonstrates various alternatives for advancement or elimination: (1) laboratory and greenhouse trials, (2) field plots, (3) demonstration areas, and (4) pilot test areas.

Laboratory and greenhouse trials. Numerous combinations of plants, spoil materials, and possible soil amendments are initially compared in laboratory and growth chamber experiments. The best alternatives are given larger trials under greenhouse conditions.

Field plots. Several alternative rehabilitation systems (combinations) are tested in field plots, which are replicated to allow for statistical analysis. The performance of each system is evaluated according to the amount of vegetative cover and forage production, as well as the effectiveness of erosion control.

Demonstration areas. The best two or three alternatives are reestablished in demonstration areas of one to four acres. These strengthen confidence in the selected alternatives, and they teach the "how-to-do-it" aspects of rehabilitation to mining personnel.

Pilot test areas. The best alternative from the demonstration areas is reestablished in pilot test areas of ten to thirty acres. These further strengthen confidence in the success of the best alternative. They also provide an opportunity for industry and land managers to do the rehabilitation job themselves, with consultation from the research staff, thus conveying information, confidence, and responsibility from the reseacher to the practitioner.

Using the scientific method to eliminate alternatives

The basis for the elimination of alternatives at each step is the scientific method, which is a format for experimentation. In any particular study, a number of hypotheses are put forth. The study contains "crucial experiments" which disprove one or more of the hypotheses, thereby lending credence to (though not proving) those hypotheses that remain. In each experiment, the scientist varies the value for one or more environmental factors and measures the response of the variable whose behavior is under study.

The use of the scientific method is illustrated in a study conducted by Knipe (1967) at the Experiment Station on the effect of temperature on the germination of alkali sacaton, an important range grass used in reclamation in the Southwest. The goals of the study were to determine if temperature had any effect on germination, and if so, what temperature caused the highest percentage of germination. To achieve this, an experiment was conducted wherein temperature was varied and germination was measured.

In the scientific method, the various possible results are anticipated in terms of hypotheses which can be selectively eliminated based on the actual findings. One possibility in Knipe's study, called the "null hypothesis," is that germination is the same at all temperatures. Another hypothesis is that there are differences in the amount of germination depending on temperature. Within the second hypothesis is a family of mutually exclusive hypotheses, each postulating that a given temperature causes significantly better germination than the other temperatures. These hypotheses are resolved by the process of elimination as the results in Fig. 4.1 are interpreted; but first the methodology requires some elaboration.

Seeds (caryopses free from all attached parts) were randomly selected from three-year-old lots and placed in batches of one hundred in a growth chamber with 100% relative humidity and with an 8-hour-light/16-hour-dark artificial light cycle. The seeds were placed in petri dishes on saturated disks of standard blue germination blotter paper. The temperatures in the growth chambers were as listed in Fig. 4.1. Four one hundred-seed replicates were subjected to each temperature regime. After 32 days, the percentage of germination for each replicate was recorded, and a mean value was calculated.

The data in Fig. 4.1 were then analyzed by using statistical tests. The first test, called "analysis of variance" (Sokal and Rohlf, 1969), indicated that there were significant differences among the percentages of germination which occurred at different temperatures. The analysis of variance (ANOVA) eliminated the null hypothesis. Another statistical test, called Duncan's multiple range procedure, was used to locate the significant differences among the germination data. This test revealed that the percentage of germination at each temperature differed significantly from the values at other temperatures, except at 50°F and 110°F, where no germination occurred.

The experiment indicated that germination depends strongly on temperature, with the highest percentage occurring at 90°F. These findings on the behavior of alkali sacaton had a high degree of confidence; they served as a basis for management recommendations and as a premise in subsequent studies of alkali sacaton, which are chronicled in the case study.

Advantages and limitations of experimentation and the scientific method

Experiments are regarded by some researchers as the only scientific tool capable of proving *causal* relationships in nature. Other scientific procedures, in which variables are measured, but not manipulated, merely imply causality. Since Knipe held constant all the variables in the growth chamber except temperature, there is no basis for disputing that temperature caused the differing percentages of germination. Such experiments simply leave little room for subjectivity or bias in the researcher's assumptions or interpretations of the results.

Another favorable aspect of experimentation is *reproducibility*. Anyone with a growth chamber and three-year-old seeds of alkali sacaton can reproduce the experimental design to confirm or refute the findings in Fig. 4.1. This may be reassuring to a reclamationist whose program depends on the germination behavior of alkali sacaton. Most other methods and studies described in this book cannot be fully reproduced, simply because they depend on the assumptions and natural conditions that prevail at the time of the research.

However, the results of experimentation apply only to a specific combination of circumstances which may never occur in nature. The experimental results describe germination with 100% relative humidity, an 8-hour-light/16-hour-dark light cycle, and whatever other environmental conditions existed in the growth chamber during the study. Thus, there is an element of uncertainty in the range of applicability of those results to field conditions, that is, in a seedbed.

Another consideration is that many phenomena of interest cannot

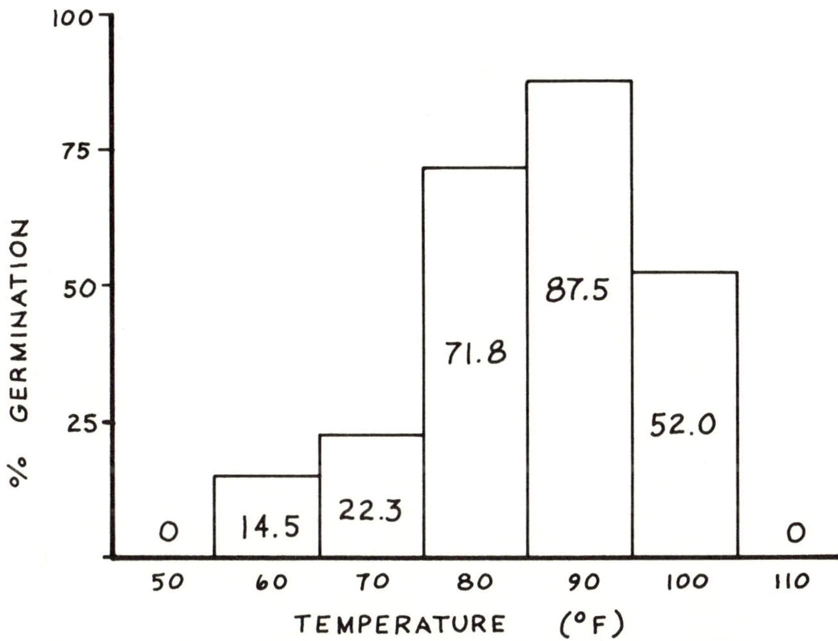

Figure 4.1. Percentages of germination of seeds of alkali sacaton at different temperatures. (From Knipe 1967.)

be studied with the degree of control achievable in a growth chamber. Most reclamation practices must eventually be tested on a large scale, such as the demonstration or pilot-test areas described above. As soon as the experiment moves outdoors, there are compromises in the reproducibility of the study and in the degree of control which the researcher has over the variables to be manipulated and those to be held constant. The reduced control under field conditions does not diminish the value of experimentation as a scientific approach. This simply demonstrates that experiments, like all scientific endeavors, often involve tradeoffs between the researcher's control of the environment and the breadth of applicability of the results.

A Case History: Development and Testing of Establishment Practices for Alkali Sacaton

Of the numerous facets of reclamation studied by the Experiment Station, the establishment of native species proved to be the paramount problem. Alkali sacaton is an important range grass in the Great Basin

Figure 4.2. Sequence of experimental tests conducted to determine the optimum combination of variables for establishing alkali sacaton on mine spoils.

Test #	Constant Temperature (°F)	Moisture Stress (atm.)	Seed Size*	Exposure to Light	Rewatering (days after planting)	
	50	0	Small	None	1	
	60	1	Large	Flash	5	
	70	4		5 min.	10	Solid line shows
	80	7		1 hour	15	best combination
1	90	10		9 hours	20	of variables.
	100	13		15 hours		
	110	16		Continuous		

Test 1 included all temperatures; test 2 used 90°F temperature only and all stresses; test 3 used 90°F, 0 moisture stress, and all seed sizes; test 4 used 90°F, 0 stress, large seeds, and all light exposures; test 5 used 90°F, 0 stress, large seeds, continuous dark conditions, and all rewatering times.

*Six sizes were used. The largest (size 1) were seeds that did not pass through a screen with 28 x 28 openings per square inch. Size 6 seeds were those which passed through the smallest (36 x 36) screen.

desertscrub biome, which is a principal vegetation type occurring in the San Juan Basin (Brown et al. 1979). A continuing project at the Experiment Station has been the testing and development of reclamation practices to ensure and maximize the establishment of this desirable perennial grass on mined land.

Laboratory, growth chamber, and greenhouse studies

Scientists at the Experiment Station conducted numerous experiments to evaluate characteristics of alkali sacaton. The effects of light, temperature, and moisture stress on germination and seedling survival were meticulously analyzed by using procedures like those in the study by Knipe (1967), described above. The findings were interpreted using "strong inference" (Platt 1964) and incorporated into subsequent experiments, leading to a continually increasing breadth and confidence of understanding of the behavior and requirements of seeds and seedlings of alkali sacaton (Aldon 1975). Figure 4.2 presents a simplified scheme of the sequence of experiments and decisions whereby we systematically increased our knowledge of this species.

This knowledge was interpreted in light of existing climatic patterns to postulate the following sequence of events required for the establishment of alkali sacaton in the San Juan Basin. Mature plants produce seeds in late summer and fall. Cold winters and dry springs do not adversely affect the seeds, but they preclude premature germination. Seeds germinate in July with the onset of late-afternoon convective thunderstorms. Storms with 25 mm or more precipitation cause flooding and sediment movement in arroyos and floodplains, transporting seeds and depositing them in saturated sediments. The seeds imbibe moisture overnight and germinate the next day, when air and surface temperatures reach 90°F. The seedling is capable of emerging through as much as 4 cm of sediment. To increase chances for seedling survival, a second thunderstorm, with 6 mm or more of precipitation, should occur within five or ten days of the first thunderstorm.

At the conclusion of laboratory and greenhouse studies, the Experiment Station arrived at the following guidelines for maximizing the germination and early survival of alkali sacaton: (1) seeds should be placed on a moist soil with a 13-mm mulch when the temperature is around 90°F; (2) agar or some other moisture-retaining colloid should be used to improve the availability of water; and (3) plants should receive supplemental irrigation if there is less than 6 mm of natural precipitation during the five days following seeding. While these measures, if strictly executed, will almost ensure successful germination and early growth, they are not necessarily practical on a large scale such as on a reclaimed

Table 4.1. Combinations of soil amendments, watering schedules, and seed placements in Treatments A through F.

Treatment	A	B	C	D	E	F
Spot saturated initially	X	X	X	X		
Agar plate (Colloidal gel)	X	X				
Mulch (Straw mulch @ 7.5 cm)	X	X	X	X	X	X
Rewater after 5 days (mm)	50	6	50	6	6	6
Seeds in depression					X	X

surface. The next step was to implement and test these guidelines in varying combinations, investing varying degrees of labor and expense, to observe their performance in the field.

Field studies

Our first field tests of methods for the establishment of alkali sacaton were at a harsh spillway site on the Rio Puerco drainage, which lies between Albuquerque, New Mexico, and the San Juan Basin. Six different treatments were tested, each composed of a different combination of soil amendments, watering schedules, and seed placement. Table 4.1 lists the combinations. Each treatment was replicated five times, with each replicate being randomly located in each of six rows. This "randomized row-plot design," illustrated in Fig. 4.3, was used to account for uncontrollable variability in the properties of the underlying soil. The random arrangement makes it unlikely that a given treatment will coincidentally be associated with a particularly favorable or unfavorable patch of soil within the study area.

The study began after the onset of summer thunderstorms and lasted for two growing seasons, during which the germination, performance, and survival of alkali sacaton were monitored. Ten large one-year-old alkali sacaton seeds were placed inside 5-cm high collars (Fig. 4.4) which were labeled A, B, C, D, E, or F, to designate the treatments in Table 4.2. Soil properties, temperature, and precipitation were recorded throughout the study to facilitate interpretation of the results.

Alkali sacaton survived and performed best in treatments C and D, where 44% of the seeds were represented by mature plants at the end of the second growing season. The results led us to eliminate several measures, such as application of agar and placement of seeds in depressions, and to recommend the following agronomic practices for further

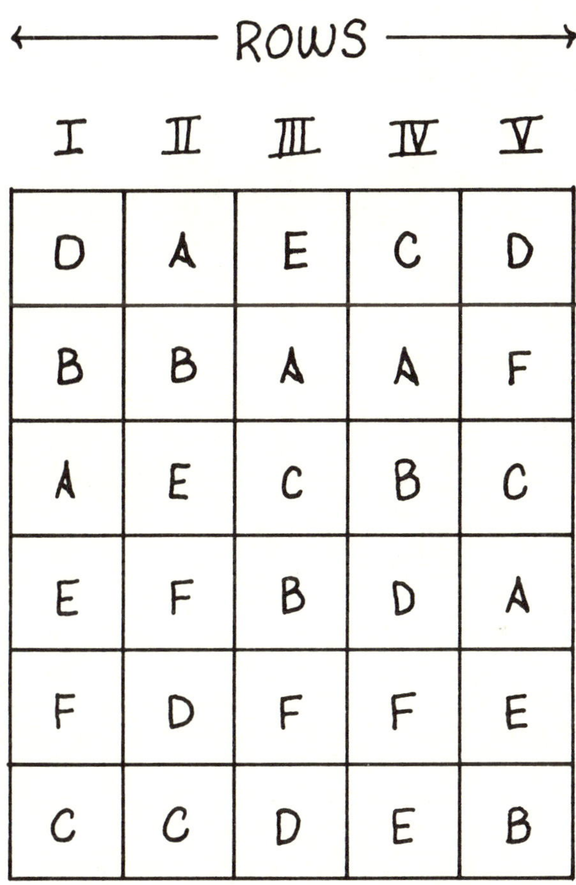

Figure 4.3. A randomized block design. Letters refer to treatments in Table 4.1. Replicates are randomly arranged in each row to account for uncontrollable spatial variability in soil properties.

Figure 4.4. Treatments in Table 4.1 applied to alkali sacaton in small collars.

consideration: (1) seed when soil moisture exceeds 13% (< 1 atm tension); (2) seed when soil temperatures will be near 90°F and when the probability for weekly precipitation is greatest (unless irrigation can be provided); (3) use large seeds which are at least one year old; 4) saturate the soil just prior to planting; and 5) cover the seeds with about 13 mm of mulch to keep them moist and in the dark.

The next step was to implement and demonstrate these agronomic practices on a large scale, in conditions as close as possible to those encountered by the practitioner of reclamation. For this purpose, two 20-acre demonstration areas were established on spoils at Mine B (on the map of the San Juan Basin in Chapter 1). The mining company furnished all the equipment and labor necessary to grade the spoils, apply amendments, apply seed, install and operate irrigation systems, and even collect field data. The surface mine is located in one of the hottest and driest parts of the San Juan Basin, which is why irrigation was incorporated into the experimental design.

The study was not only conducted on a very large scale, but it encompassed many different aspects of reclamation, including surface amend-

Table 4.2.   Physical and chemical characteristics of the three spoil types at Mine B.

Properties	Grayish Brown Sandy Loam	Gray Stoney Clay	Grayish Brown Shaley Clay
% Sand	75.1	22.6	19.4
% Silt	8.4	19.6	35.6
% Clay	16.5	57.8	45.0
Sodium adsorption ratio (SAR)	8.8	27.7	5.6
Electrical conductivity (mmhos/cm)	7.7	13.3	7.0
pH	7.3	7.3	7.4
Na (meq/1)	39.1	106.5	27.8
Ca + Mg (meq/1)	39.2	29.7	50.0

ments, seed mixes, and rates and methods of irrigation. Aldon et al. (1976) provide an overview of this project. The design of the demonstration plots and some aspects pertaining to alkali sacaton are discussed herein.

The spoils at Mine B were graded to a gently rolling topography with slopes generally less than 6%. After grading, three different spoil types were identified, mapped, and analyzed for the properties in Table 4.2. The 20-acre plots were located on the graded spoils so that the three spoil types were about equally represented. Irrigation water was applied with sprinklers on one plot and drip-emitters on the other plot.

Figure 4.5 illustrates the layout of a demonstration plot on the three underlying spoil types. Three spoil amendments were applied in 5-m to 8-m wide strips across the plots. The amendments included: (1) sandy topsoil, salvaged from native habitat, applied to a depth of 15 to 25 cm; (2) bottom ash, from the boilers of an adjacent coal-fired power plant, applied to a depth of 15 to 25 cm, and then plowed and disked; and (3) gypsum ($CaSO_4 \cdot 2H_2O$), applied at a rate of 7.75 tons per acre, and then disked. The location of these strips was determined by drawing random numbers. Different seed-mixes and irrigation rates were also tested at various locations in each plot, although these are not illustrated because a diagram including all treatments is very complicated.

The specific data and statistical analyses are presented in Aldon et al. (1976). Alkali sacaton was generally the most successful species in the seed mixes with regard to emergence and survival. Emergence was substantially higher on the grayish-brown sandy loam than on the oth-

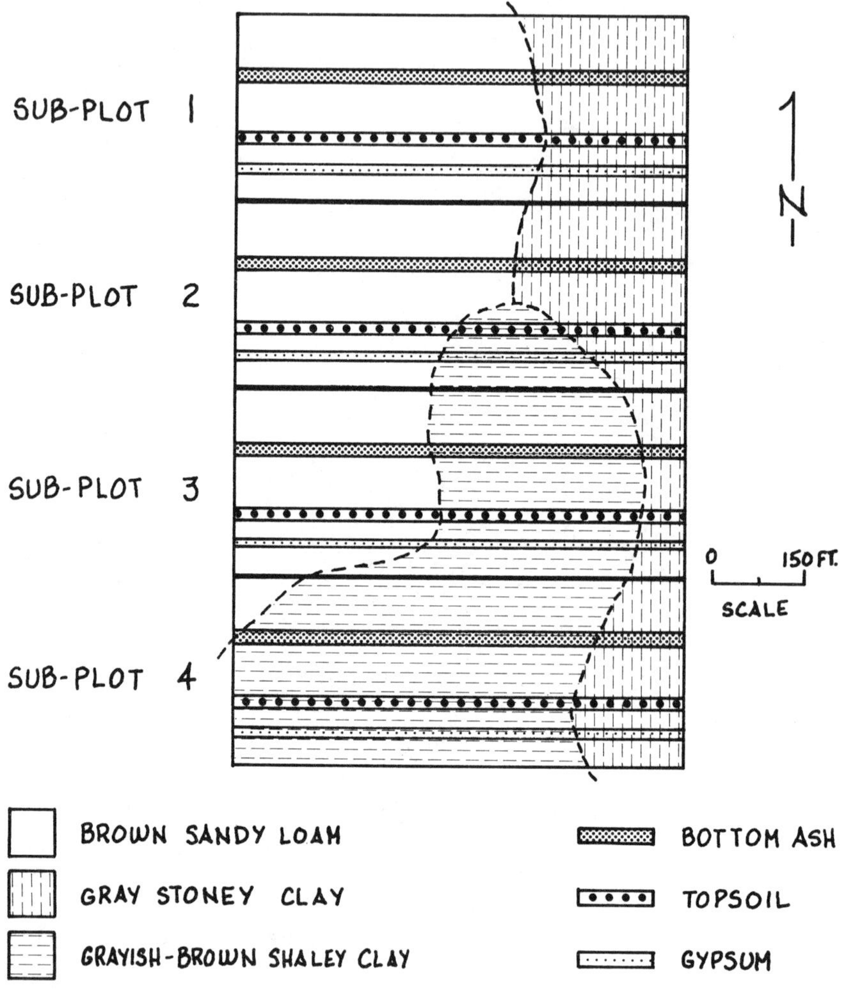

Figure 4.5. Layout of demonstration area at Mine B.

er two spoil types. Survival of alkali sacaton after two growing seasons was better on topsoil and bottom ash than on gypsum or unamended spoil. The average density under sprinkler irrigation exceeded nine plants per square meter, which compares favorably to densities observed in natural habitats (Aldon, 1975).

Future Activities of the Experiment Station

Much of the information from the Experiment Station's research program has been incorporated into actual reclamation policies and practices at surface coal mines in the San Juan Basin and elsewhere. The establishment methods described above have been adapted to fit local conditions and have resulted in good stands of perennial grasses and shrubs on reclaimed surfaces.

However, many questions still remain about reclamation in arid and semiarid habitats. Will reseeded habitats have adequate reproduction, thus becoming self-propagating, stable ecosystems? What plant densities and species compositions constitute an adequate stand relative to the pre-mining condition of the range? Will salt migration in saline and sodic spoils eventually degrade the quality of the surface material and reduce vegetative cover? How should reclaimed habitats be managed?

To address these questions regarding the long-term durability of revegetated habitats, the Experiment Station is currently studying factors related to plant succession, site productivity, and soil development. For instance, we have one project under way on the bacterial and fungal populations, as well as the decomposition rates, in mine soils which have been subjected to a variety of reclamation treatments. Another project is tracking the long-term behavior of water and salts on relatively old reclaimed surfaces.

This research will increase our understanding of the long-term aspects of reclamation, so that reseeded habitats, once established, will become durable and productive vegetation communities.

References

Aldon, E. F. 1975. Establishing alkali sacaton on harsh sites in the Southwest. Journal of Range Management 28:129-132.

Aldon, E. F., H. W. Springfield, and W. E. Sowards. 1976. Demonstration tests of two irrigation systems for plant establishment on coal mine spoils. *In* Fourth symposium on surface mining and reclamation, 201-214. NCA/BCR Coal Conference and Exposition 3, Louisville, KY. Washington, DC: National Coal Association.

Brown, D. E., C. H. Lowe, and C. P. Pase. 1979. A digitized classification system for the biotic communities of North America, with community (series) and association examples for the Southwest. Journal of the Arizona-Nevada Academy of Science, 14 (suppl. 1): 1-16.

Knipe, O. D. 1967. Influence of temperature on the germination of some range grasses. Journal of Range Management 20:298-299.

Platt, J. R. 1964. Strong inference. Science 146: 347-353.

Sokal, R. R., and F. J. Rolfe. 1969. Biometry. San Francisco, CA: W. H. Freeman and Company.

U.S. Department of Agriculture. 1983. Agricultural handbook 453. Washington, DC: Government Printing Office.

5

Understanding Reclamation with Models

Charles C. Reith

The Need for Models in Reclamation Science

Many acres of mined land have received some degree of rehabilitation management over the last fifteen years. The results of these reclamation efforts include many apparent successes—landscapes with diverse plant communities which are reproducing and stabilizing soil against erosion. These reclamation efforts and their outcomes should constitute a valuable source of information, especially at older mines where the durability of the vegetation and the relatively long-term stability of the reclaimed surface can be observed.

However, there are difficulties in using the information represented by surfaces reclaimed in the past. One problem is incomplete data in the management history of a reclaimed surface. Evaluations of management strategies are difficult when data such as the following are unrecorded or imprecisely documented for a reclaimed surface: (1) rates of application of irrigation, mulch, fertilizer, topsoil, and other surface amendments; (2) mechanical treatments of the mine soil, such as ripping, disking, or furrowing; or (3) rates of application and compositions of seed-mixes. Even when such data are relatively complete, a scientist may lack confidence in the precision of execution of the management program, at least compared to the precision usually associated with experimental procedures.

Another problem is that management characteristics on reclaimed surfaces may not be controlled so as to allow evaluations of single factors. For instance, it is difficult to assess the effectiveness of using gypsum to alleviate sodicity if one reclaimed surface was treated with gypsum and some other surface amendment—say, bottom ash—whereas a similar surface received neither gypsum nor bottom ash. If the former surface produces more vegetation, one cannot ascertain whether the improvement is due to the gypsum, bottom ash, or the combined action of both. To achieve a high degree of confidence in assessing the effect of one factor on reclamation, scientists conduct experiments (see Chapter 4) wherein all variables are controlled except the variable under study. Some scientists regard experiments as essential for establishing "scientific proof" of any relationships or phenomena in nature.

Modeling is one way to overcome the potentially confusing variability associated with surfaces reclaimed in the past. Models simplify environmental phenomena, such as reclamation, by eliminating those aspects which seem relatively unimportant. The result of modeling, after simplifications and assumptions, is a "system" of interacting components and processes. The behavior of such a system can then be analyzed and evaluated with respect to its fidelity to nature (the degree to which the system's behavior corresponds to the behavior of the natural phenomenon being modeled). An important goal of modeling is to make predictions. In the case of reclamation, this goal means predicting reclamation potential, that is, predicting the degree of revegetation and stabilization which will result from a given combination of environmental conditions and reclamation practices.

The Philosophy and Jargon of Modeling

Modeling is the principal method behind "systems analysis," a scientific approach which is motivated in part by the following observation.

> No substantial part of the universe is so simple that it can be grasped and controlled without abstraction. Abstraction consists in replacing the part of the universe under consideration by a model of similar but simpler construction. Models, formal or intellectual on one hand, or material on the other, are thus a central necessity of scientific procedure. (Rosenblueth and Wiener 1945)

Thus systems analysis simplifies natural phenomena to a level where the behavior of the phenomena can be analyzed or simulated, and ultimately predicted. An excellent book by Jeffers (1978) precisely defines the role of modeling in the overall scheme of systems analysis and rigorously classifies models according to their objectives, assumptions, and mathematical rationales. This chapter provides a more general explanation and classification of modeling as a tool for understanding and predicting the events and relationships in reclamation.

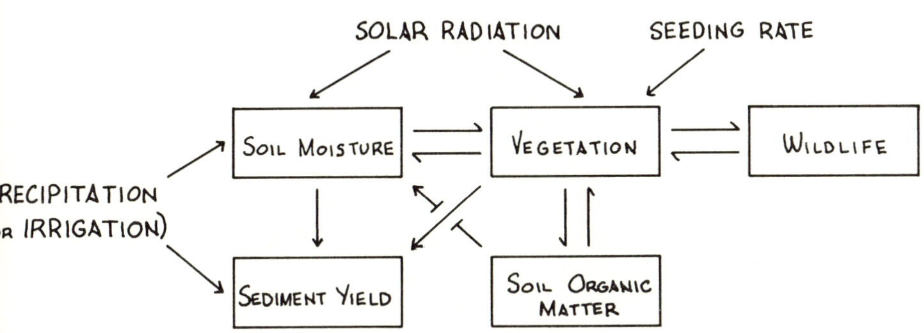

Figure 5.1. A simple model of components, processes, and external factors at a reclaimed surface.

Components and processes

A model represents a natural phenomenon as a network of *components* and *processes,* terms which will be defined in reference to Fig. 5.1, a highly simplified model of a reclaimed surface. The components are the boxes; these describe various aspects of a reclaimed surface that we can observe and measure. A component is generally described according to its *state,* which is the condition of that component. The central component in Fig. 5.1 is vegetation. The state of the vegetation is the amount at any given time, expressed in biomass, coverage, and so forth. If the model is quantitative, the state of vegetation is expressed with numbers, for example, kilograms/hectare or percentage of basal cover. If the model is qualitative or conceptual, the state of the vegetation is described as "good" versus "bad," or as "adequate" versus "inadequate."

The other components in Fig. 5.1 are likewise described according to their states: wildlife according to abundance or diversity, soil moisture according to gravimetric content or potential, and sediment yield according to some measure of erosion rate. The states of these components are the result of (1) the influence of factors external to the system and (2) the processes indicated by the arrows between components. Precipitation and solar radiation are external factors because they are unaffected by anything in the system, which is realistic because soils, vegetation, and wildlife do not influence climate (except in the case of large-scale efforts to reclaim desertified land). Rehabilitation practices such as seed-

ing rate and irrigation are also external if system components do not
influence them. However, if irrigation is adjusted in response to the
amount of vegetation or soil moisture, then its role in the system shifts
from being an external factor to a component.

Processes in the model, which are represented by arrows, are those
events and relationships that determine the states of the components.
The most important process in Fig. 5.1 is the change in the amount of
vegetation which results from growth or dieback. This process initially
depends upon seeding rate, and then upon external factors, such as solar
radiation, and upon system components such as soil moisture and wildlife. These dependencies are illustrated by the arrows which point to
vegetation from the sources of influence.

Most of the arrows between components are two-way, reflecting the
interdependency which is commonplace in nature. Soil water enourages
plant growth, which in turn depletes soil water; this is a *negative
feedback* relationship because an increase in one component (vegetation) causes a decrease in the other (soil water). Other relationships are
positive feedback, where an increase or decrease in one component moti-
vates a parallel response in the other. For instance, organic matter in
the soil results in desirable water relations and aeration, thus favoring
plant growth. At the same time, the roots and litter of plants add organic matter to the soil.

The disputability of models

The relationships in this model of reclamation, and in most models,
are disputable. A model is simply the modeler's interpretation of what
happens in nature, an interpretation which may summarize a library of
experimental data or result from a moment of speculation. The states
and processes in Fig. 5.1 describe one interpretation of what is important in reclamation. Components or relationships which are not illustrated in the model are not necessarily absent in the natural phenomenon
being modeled, but they are deemed relatively unimportant.

For instance, erosion undoubtably influences vegetation by undermin-
ing and exposing the roots of upslope plants and burying downslope
plants under sediment. Then why is there no arrow pointing from the
component for sediment yield to the component for vegetation? The
modeler might respond by arguing that such effect is really quite minor
compared to the other factors influencing vegetation; thus, including
the process would unnecessarily complicate the model. Eventually, however, some modeler will be dissatisfied with this assumption, and the
effect of erosion will be included in a new, more sophisicated model

with some small, but possibly important, improvement in the model's performance.

The disputability of a model is, in a sense, essential to the fulfillment of the model's principal objective: to illustrate a natural pheomenon in such a fashion as to provoke and enable further study. The question posed above is but one of many which a rough model like Fig. 5.1 should provoke. At the same time, the model reduces the complex phenomenon of reclamation into a network of simpler phenomena which can be studied by using experiments and other tools of scientific research, nearly all of which begin with a precise definition of the problem.

Types of models

There are several kinds of models, but it is difficult to present a classification scheme because the types of models are not entirely distinct from each other and because any given model may be a hybrid of several different model types. My approach to model classification is to present those terms in the literature which may pertain to models in reclamation science. These terms describe models according to their properties and objectives, which are often linked. The most encompassing classification terms are listed first, and dichotomous classifications are noted.

Conceptual. Every model is initially conceptual, in that natural phenomena are conceptualized into a format which can be communicated via words or pictures. A conceptual model may be highly sophisticated and detailed, incorporating numerous subtle relationships, but it will never generate precise predictions because states are not quantified and processes are not translated into equations. Conceptual models fulfill the principal goal of simplifying nature into a format that can be studied; they also provide a framework for mathematical models.

Mathematical. Few modeling efforts conclude at the conceptual stage because scientists wish to add precision to their description of nature; they wish to describe states in terms of numbers instead of words such as "high," "moderate," or "low." In mathematical models, the states of all components are represented by numerical values or *magnitudes.* These magnitudes change in response to the arrows which link components and which represent mathematical equations. Thus, a mathematical model is essentially a series of equations which can be solved to determine the magnitudes of various states. These magnitudes may be interpreted to be precise predictions of the behavior of the system being modeled.

Simulation. A simulation model attempts to simulate the behavior of a system through *time,* hence it is also called a dynamic model. Data are not necessarily required. For instance, Fig. 5.1 becomes a simulation model if we assign starting values (states) to the four components and then construct mathematical equations to describe how the states change over time in response to each other and to the external factors (rain and sun). The starting values and equations may be based on data or speculation. The model is then validated by comparing its predictions to actual values from the system being modeled. Computers are invariably used because mathematical equations are repetitively calculated to simulate the passage of time in the system. A simulation model is the only scientific tool which enables precise, long-term predictions.

Analytical.   Analytic models are used by scientists to explore and explain data which they have collected. Analytical models organize data into patterns and distributions which reveal relationships among the factors being studied. Statistical techniques such as regression and analysis of variance are tools for analytical modeling.

Deterministic versus stochastic.   These terms describe an important distinction in the behavior of mathematical models, particularly simulation models. A simulation model is deterministic if it consists strictly of mathematical equations such as those described above. For any given set of starting conditions, the model generates the same predictions over time. However, nature has random (stochastic) elements such as rain and sun; thus, it is somewhat unrealistic for the modeler to control these variables as in deterministic models.

The equations in stochastic models include probability functions (for example, "random number generators") which introduce a certain degree of randomness into the behavior of the model. These probability functions are often based upon mathematical distributions such as the normal or Poisson distributions, which are widely used to study natural systems. Because of the random factor, stochastic models produce different results on every run. Stochastic simulation models are generally repetitively executed from the same starting conditions, so that the range and distribution of predictions can be observed.

Limitations of models

So far, models have been discussed with reference to their ability to organize and explain information which is not conducive to other forms of scientific study, a valuable attribute considering that most reclamation has been conducted with little regard toward future analyses. However, the value of modeling is limited by the degree of confidence one can have in (1) the validity of the assumptions in models, (2) the nature of the relationships identified or simulated by models, and (3) the predictions generated by models.

A model is, by definition, a simplification or approximation of a real system. Modeling begins with carefully specified assumptions as to which elements of the system will be disregarded. These factors may be absent because of a lack of sufficient data or knowledge to account intelligently for their behavior in the model. The modeler has no choice but to assume that such factors are unimportant, even though the accuracy of the model depends on the validity of the assumptions. If a simulation model behaves nothing like the system being modeled, then the modeler checks the assumptions, as well as the equations which drive the model, in search of inaccuracy. If an analytical model generates relationships which seem implausible in the system under study, then one of the assumptions may be invalid.

In any case, the relationships in a model are based solely upon the modeler's interpretation of the natural system (in simulation models) or upon the patterns revealed by statistical analysis (in analytical models). Regardless of the apparent accuracy with which the model describes and predicts nature, the relationships in the model are not necessarily *causative*. If Component-A depends on Component-B according to an equation in the model, and if the model behaves accurately, this is not *proof* that changes in Component-B cause changes in Component-A, although the model provides a basis for suspecting a causative relationship and perhaps for initiating an experimental study to verify causality.

Confidence in the predictive accuracy of a simulation model depends upon the thoroughness and range of validation. As mentioned earlier, validation entails comparisons of the model's predictions to actual values observed in nature. Confidence grows as the model seems to correspond to the behavior of the modeled system over a broader range of circumstances, although confidence is always greater within the range of validation than beyond it. For instance, suppose the model in Fig. 5.1 accurately predicts the amount of vegetation with 10 cm and with 40 cm of annual precipitation. We should be more confident in the model's projection of the amount of vegetation with 25 cm than with 100 cm. Unfortunately, models are frequently and understandably used to predict the results of extreme or unusual circumstances which fall outside the constructs of the model; prolonged drought, say, in the case of reclamation.

Charles C. Reith

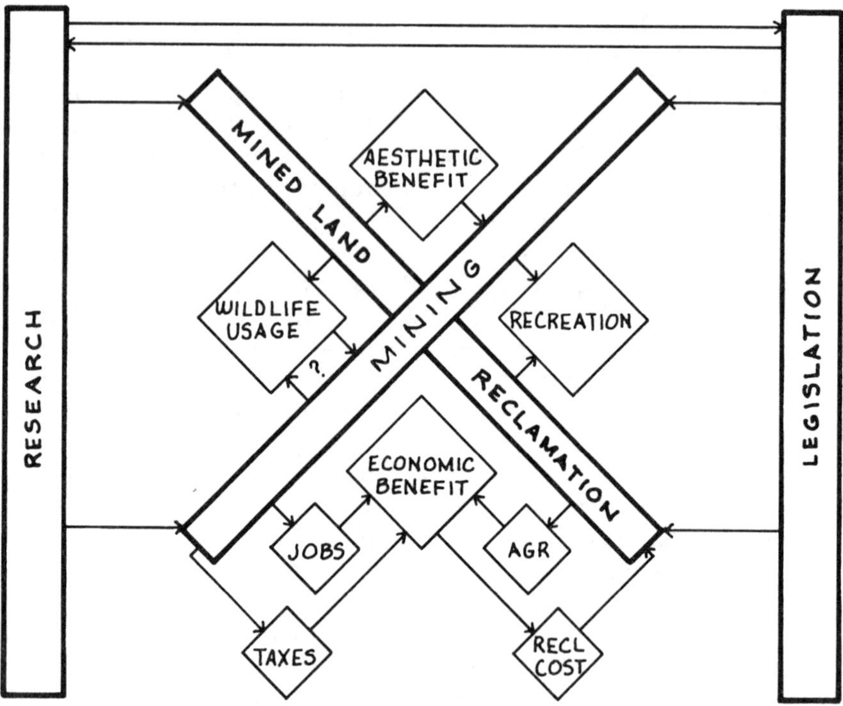

Figure 5.2. A conceptual model illustrating the context of land reclamation in a network of social and environmental factors. (From Wali 1975.)

General applications of modeling to reclamation

Despite the potential utility mentioned above, models have not been widely applied to reclamation. Instead, research has emphasized specific aspects of reclamation, testing the effectiveness and applicability of various management practices and surface amendments. Most modeling efforts have examined reclamation from a broader perspective, defining and explaining the relationships among the numerous components and processes. A few of these modeling projects are briefly described here to illustrate the diversity of applications of models to reclamation.

Conceptual models. Wali (1975) viewed reclamation from the broadest perspective by using a conceptual model (Fig. 5.2) to illustrate the context of coal mine reclamation relative to other issues and processes in energy development. The model supports Wali's point that the cost of reclamation should depend on a network of factors related to the val-

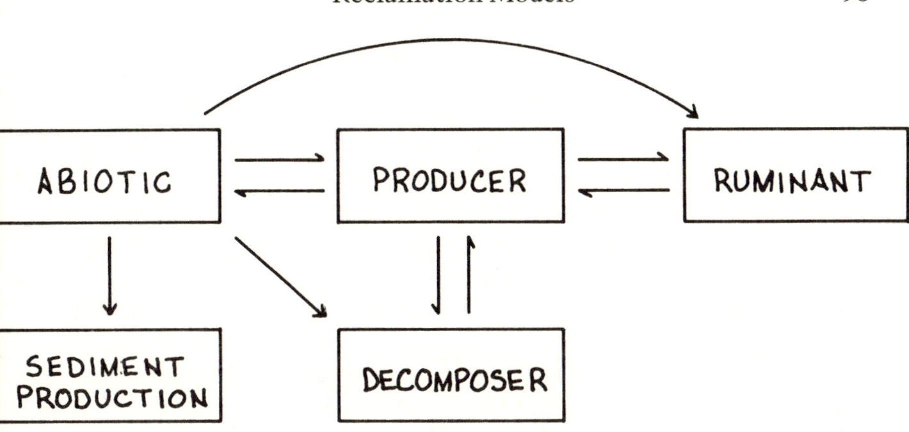

Figure 5.3. Submodels of the large simulation model of reclamation at a grassland ecosystem in Montana. (From Parton et al. 1979.)

ue of energy and land, and to the social and environmental consequences, should reclamation not be accomplished. In the same paper, Wali presents another conceptual model, in the form of a long and detailed flowchart, that illustrates the hierarchy of variables which should be studied in attempting to improve our knowledge of reclamation in arid or semiarid climates where spoils may be sodic, saline, or clay rich. This model is essentially a framework for reclamation research.

Simulation model.   An ambitious simulation model was constructed by the Natural Resource Ecology Laboratory to describe and test reclamation practices on a grassland ecosystem in eastern Montana. The model is presented in detail in Ellis and Parton (1978) and is summarized in Parton, Ellis, and Swift (1979). The deterministic model consists of five submodels (see Fig. 5.3), each of which contains ten or more components linked by various processes. The mathematical equations which drive the model are executed using SIMPCOM, a FORTRAN-like computer language. The model predicts soil moisture content and live plant-shoot biomass, among other things. The performance of the model was verified by comparing two-year-long records of values for these parameters, as predicted by the simulation model, to the actual records observed in the natural system. Although the simulation behaved closely to the natural system, the authors did not consider this true validation because two data sets should be used: one to develop the model and a second to test the model. The same data set was used for both development and testing of the model because only one data set was available. The authors point out, however, that the model generates reasonable predictions in response to manipulations of the starting conditions and the external factors which influence the simulated ecosystem.

The reclamation model was used to conduct "simulation experiments" in which various elements (components, processes, starting conditions, or external factors) were manipulated in order to anticipate the effects of different management alternatives. Figure 5.4 illustrates the model's predictions in response to various combinations of topsoil, fertilizer, and irrigation.

Analytical models. The most ambitious effort to predict reclamation potential over a geographic area involved the use of analytical models. Packer et al. (1982), from the Intermountain Forest and Range Experiment Station (U.S. Forest Service), collected vegetation, soils, and climatic data from reclaimed surfaces and adjacent undisturbed surfaces at twenty-eight coal mines in the Northern Great Plains states. The authors searched for relationships among the environmental parameters in this large database by using multiple regression, a statistical modeling technique which will be explained more fully in the upcoming case study. The analysis identified three parameters which seemed to be most important in determining the forage production of both introduced and native grass species on reclaimed surfaces: (1) age since planting, (2) annual precipitation, and (3) length of the growing season. The mathematical relationships among these parameters are expressed in equations which, theoretically, enable predictions of forage production at any location where the annual precipitation and length of growing season are known. These predictions may then be adjusted according to anticipated applications of topsoil, irrigation, fertilizer, and so forth.

A Case Study: Using Analytical Models
to Understand Reclamation in the San Juan Basin

A somewhat detailed account of a modeling study in the San Juan Basin (Reith and Potter 1983) will now be presented to illustrate (1) the methodology used to assemble a database for analytical modeling, (2) the assumptions behind the models, and (3) the results and interpretations of the models. The study addressed the following questions:

—Which environmental factors have been important in determining the relative success of past reclamation at surface mines in the San Juan Basin?

—Which factors are important at all mines in the Basin and which are important only at specific sites?

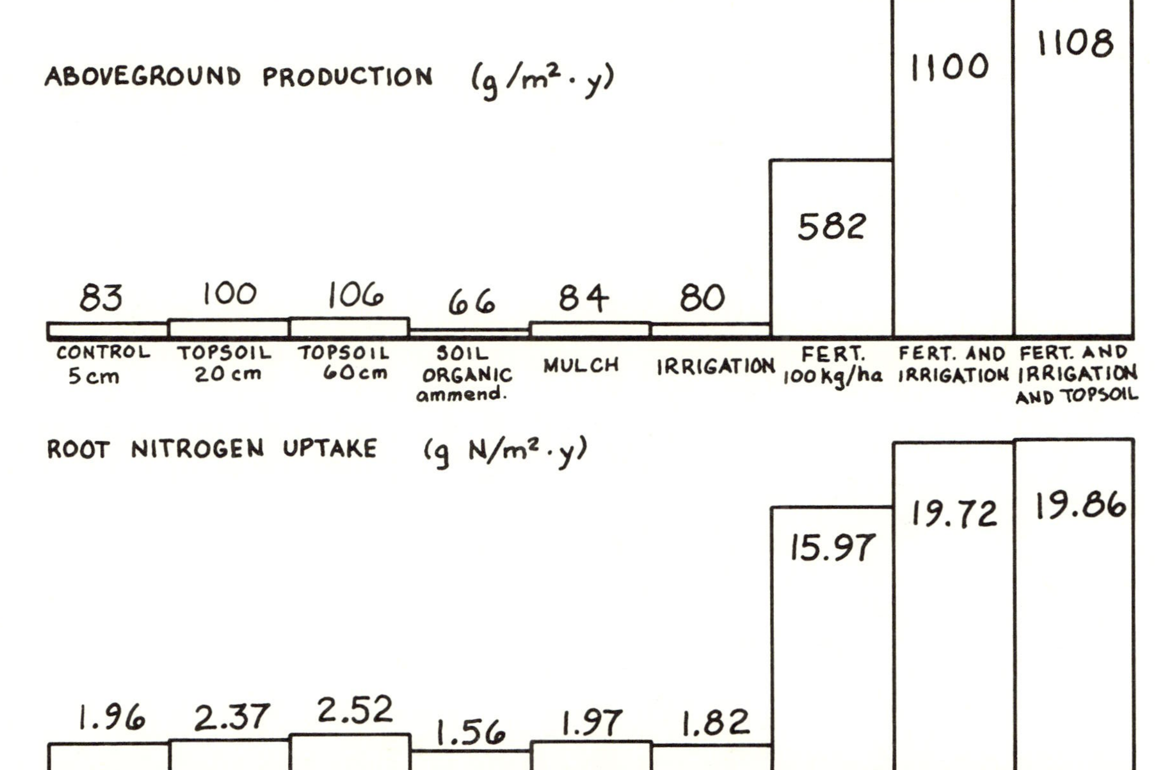

Figure 5.4. Predictions from a simulation model of a reclaimed ecosystem. Height of bars indicates response of two system parameters to different starting conditions. (Adapted from Parton et al. 1979.)

The study began with the selection of four surface coal mines at different locations in the San Juan Basin (see map in Chapter 1), representing much of the range of environmental variability present within the Basin. Each of these mines had many reclaimed surfaces of different ages, with differing management histories, and with different results, in terms of the apparent success of revegetation and stabilization. The fundamental problem was as stated in the introduction to this chapter: How can this potential source of information be analyzed and explained to improve our understanding of reclamation in the San Juan Basin?

The basic approach was to use multiple regression models to explain the variation in reclamation success among as many reclaimed locations as could be measured at each mine, with the locations representing the full spectrum of successes and failures in revegetation and stabilization. Between twenty and fifty-two locations were sampled at each mine. Fig. 5.5 lists the factors which were measured: At the top are *predictor variables*, values describing soil, topographic, and management characteristics which may influence the outcome of reclamation at each site; and at the bottom are *outcome variables*, values describing the amount of revegetation and erosion, hence serving as indices of reclamation success. For each outcome variable at each mine, a multiple regression model identified and ranked those predictor variables which were responsible for the variation in the outcome variable.

Data collection

Before the models are presented and interpreted, some details of the methodology must be discussed to explain how study sites were selected, delineated, and characterized with respect to their "reclamation success." At each mine, we searched reclaimed surfaces for small areas (about thirty square meters) of relatively uniform revegetation, specifically of perennial grasses. Perennial grasses served as our index of reclamation success because their establishment was critical to the revegetation and post-mining land-use objectives at all four mines. Sample sites were selected to *maximize* the variability in the amount of revegetation at the twenty and fifty-two study sites at each mine because the greater the variability in the outcome parameter in a model, the greater will be the variability (and the better the resolution) among the model's predictors.

The parameters in Fig. 5.5 were measured using the sampling scheme portrayed in Fig. 5.6. The biomass of perennial grasses and the volume of sediment removed by rills were estimated in each of the eight quadrats, and then averaged for the study site. The frequency of perennial grasses was calculated as the percentage of quadrats with some perennial grasses. Soil properties were analyzed from a 2-kg sample of the

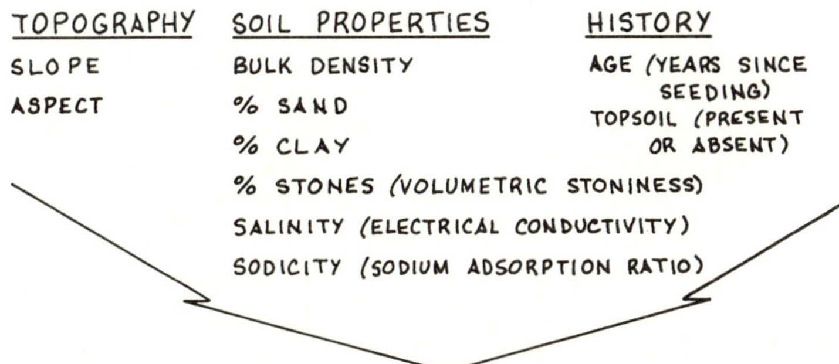

Figure 5.5. Environmental factors (predictors) and outcomes of reclamation in the case study. For each outcome at each mine, a multiple regression model identifies those predictors which are most important. (Details on measurement of parameters are in Reith and Potter 1983.)

surface material from the top 15 cm at the center of the study site. Slope, aspect, age, and the presence or absence of topsoil were also recorded for each site. The topsoil variable indicates whether or not native soil was salvaged prior to mining and then redistributed (at considerable expense) on the graded surface. Since the presence versus absence of topsoil is not a numerical value, topsoil is a "dummy variable" and is assigned a "0" if absent and a "1" if present; multiple regression will handle only one or two such dummy variables (Sokal and Rolfe 1969). More details about the actual measurement of the variables in Fig. 5.5 are in the final report for the project (Reith and Potter 1983).

Multiple linear regression and its assumptions

Multiple linear regression is an extension of linear regression, a common statistical procedure wherein the relationship between two variables is described by a correlation coefficient (r), which depends on the strength of the relationship, and by a linear equation (y = mx + b). Multiple linear regression describes the relationship between several predictor variables and an outcome variable in terms of a multiple correlation coefficient (R) and a linear equation with more variables (y = m1x1 + m2x2 + m3x3 . . . + b). In both cases, correlation coefficients are relatively high if linear relationships exist between predictor and outcome variables.

The principal assumption behind all linear regression techniques is that the participating variables are approximately normally distributed (that is, in rough correspondence to a "bell curve"; see Chapter 9 for more discussion of the normal distribution). Multiple regression tolerates a certain degree of departure from this assumption (for instance, the dummy variable mentioned above). However, confidence in a regression model decreases with increasing departures from the normality assumption, so users of multiple regression should minimize the number of non-normal variables in their models. The values for some kinds of data can be mathematically transformed to approximate normality. For instance, the percentage-variables in Fig. 5.5 (frequency, stoniness, and sand- and clay-content) were transformed by taking the arcsine of the square root, a standard transformation (Sokal and Rolfe 1969).

Another assumption in multiple regression involves sample size. Minimum sample size is not regulated by any specific guidelines, but by the fact that models are more likely to contain spurious relationships as the ratio falls between the number of samples and the number of predictor variables in the model. A spurious relationship is one that exists in the model, but not in nature; thus, it can be harmfully deceptive if considered a valid scientific finding and used for management recommendations. The rule of thumb is for models to have about ten times as many samples as variables. Models for one mine in the case study are based on only twenty samples, but identify more than two important variables; these must be interpreted with special caution since the rule of thumb for sample size was violated.

Finally, multiple regression requires independence among predictor variables, an assumption which is hardly ever fully satisfied in ecological studies because of the inherent interdependence among natural processes and entities. However, the modeler should avoid combinations of predictors which are very highly correlated to each other. For instance, the percentage of silt is absent from Fig. 5.5 because the percentage of clay and the percentage of sand were both included, and the three textural classes in combination would have a correlation of 1.0, which would violate the independence assumption. Still, modelers must tolerate some degree of correlation among the predictor variables. The bottom line with this and any assumption is to minimize violations and to qualify

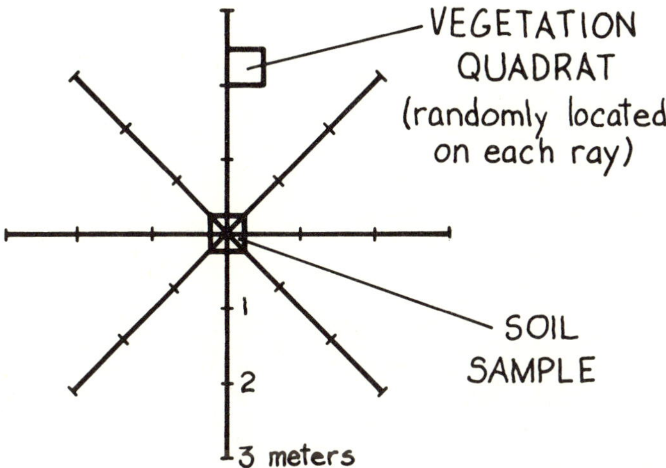

Figure 5.6. Sampling scheme for measurement of parameters in Figure 5.5.

interpretations with a recognition of the possible effects of the violations on a model's composition.

The models and interpretations

Figure 5.7 presents twelve models which explain the variation in three different outcome variables (rows) at four different mines (columns). The results of analytical modeling can be expressed in a variety of formats. Figure 5.7 includes and emphasizes those ingredients of the models which are most important in this case study: (1) the explanatory power and statistical significance; (2) the determinants, in order of importance; and (3) the direction of correlation between the determinants and outcomes. Following is a brief guide to the interpretation of each model.

R-square (R^2): This number is the square of the multiple correlation coefficient and is called the *coefficient of determination*. It indicates the degree to which the determinants in the model explain the variation in the outcome, specifically the percentage of variation which is explained. For instance, in the upper left-hand model, 32.2% of the variation in the amount of biomass at Mine A is explained by the factors in the model.

Significance level (p): This number indicates statistical significance, specifically the probability that the relationships in the model do not

OUTCOMES ⟱

	MINE A	MINE B	MINE C	MINE D
BIOMASS	SAMPLES (48)	(52)	(52)	(20)
	R^2 .322	.244	.310	.622
	P .0055	.0036	.0035	.0108
Determinants	AGE (-)	% SAND (+)	BULK DEN. (-)	SALINITY (-)
	% SAND (-)	BULK DEN. (-)	% SAND (+)	% STONE (+)
	% CLAY (-)	TOPSOIL (-)	AGE (+)	% SAND (-)
	SODICITY (-)		SLOPE (-)	AGE (+)
	SALINITY (-)		% STONE (-)	BULK DEN. (-)
FREQUENCY	R^2 .513	.425	.397	.665
	P .0001	.0001	.0001	.0130
Determinants	TOPSOIL (+)	% SAND (+)	% SAND (+)	SODICITY (-)
	SODICITY (+)	BULK DEN. (-)	ASPECT (+)	% SAND (-)
	% SAND (-)	SLOPE (-)	BULK DEN. (-)	BULK DEN. (-)
	SLOPE (+)		SLOPE (-)	AGE (+)
	ASPECT (+)			SLOPE (-)
	% STONE (-)			ASPECT (-)
RILLS	R^2 .174	.187	.395	
	P .0400	.0062	.0006	No
Determinants	BULK DEN. (-)	SLOPE (+)	BULK DEN. (+)	Significant
	ASPECT (+)	% SAND (-)	ASPECT (-)	Model
	% CLAY (-)		SLOPE (+)	
			AGE (+)	
			SALINITY (+)	
			% STONE (+)	

Figure 5.7. Models for the three outcomes at the four mines. Determinants are listed in order of importance.

really exist in nature. Any value below 0.05 is considered acceptable, but the smaller the p-value, the greater confidence one can have in the validity of the model.

Determinants: The determinants in each model are those from Fig. 5.5 which are most important in explaining the variation in the outcome. The number of determinants was selected by using a *canned* program called "stepwise" (Statistical Analysis System 1979). (Canned programs are professionally written software packages that are generally available at large computer facilities and sold as scaled-down versions for smaller systems.) Stepwise seeks and prints the best possible multiple regression model (highest R^2) for each number of determinants. The result is the curve in Fig. 5.8, which illustrates the increase in R^2 as variables are added to the model. The final models, which are presented in Fig. 5.7, are those which have the highest R^2 with the least

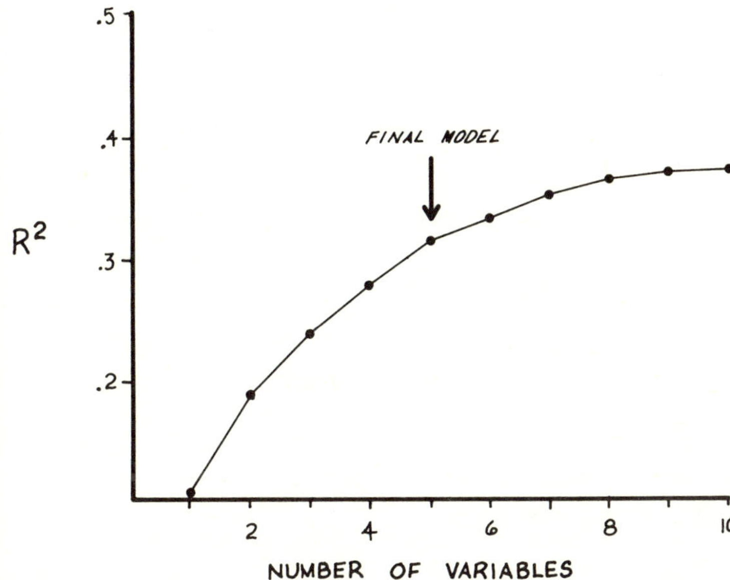

Figure 5.8. R^2 increases as more variables are added to multiple regression models. The arrow indicates the number of variables in final model.

number of determinants, and thus the highest power to explain the outcome, but with the least complexity. This decision, represented by the arrow in Fig. 5.8, is made by the modeler, who seeks the simplest, most powerful combination of determinants.

The determinants are listed in decreasing order of their importance in each model. This is based on a significance value which is printed by the computer, but not portrayed in Fig. 5.7, and which relates to the individual contribution of each determinant to the overall model. The boldface determinants are those whose individual contribution to each model is significant; interpretations should emphasize these variables.

A final, but important, symbol in Fig. 5.7 is the plus or minus sign beside each determinant that indicates the direction of correlation with the outcome.

Interpretation.   A somewhat detailed interpretation of the models at Mine C will be given to illustrate how the modeler analyzes and explains the relationships to arrive at recommendations. The model for frequency has a higher R^2 and lower p-value than does the model for biomass, suggesting that the determinants in this study are more effective in explaining the presence-versus-absence of perennial grasses than the amount of perennial grasses. This was true at all four mines. Bulk density is a negative correlate in both revegetation models, suggesting that excess soil compaction interferes with the establishment and growth of perennial grasses. Compacted soils may lack structure required for good infiltration of water and aeration of roots. It is true, however, that vegetation gradually reduces soil bulk density by the action of roots and the addition of organic matter, so there is some ambiguity as to which factor is the determinant and which is the outcome.

Sand is also a positive correlate in both revegetation models, suggesting that perennial grasses reproduce and grow best in sandy materials at Mine C. Sand may be especially beneficial in arid soils because it allows for better infiltration and withholds less moisture via matric-retention than does clay, an advantage where rainfall may be infrequent but torrential. The positive correlation of age in the biomass model indicates that there was more perennial grass on older surfaces, and the positive correlation of aspect in the frequency model indicates that there were fewer bare patches on north- and east-facing slopes than on south- and west-facing slopes, which corresponds to north-south effects which are common in nature.

Bulk density was the only significant correlate in the model explaining rills. The positive correlation suggests that erosion was mostly on compacted soils, possibly because these soils resist infiltration of water, generating more runoff which removes sediment. Also, since compaction inhibited revegetation, there was less perennial grass to stabilize compacted soils. The rill model includes a relatively long list of non-significant variables; evidently, none of these were singularly important, but they influenced erosion in combination. The model for rills at Mine C was the only erosion model with a relatively high R^2, possibly because that mine was the only one with much topographic diversity, that is, with many slopes exceeding ten degrees.

Recommendations. The recommendations based on the models at Mine C are fairly straightforward: (1) keep bulk densities low by encouraging soil structure (crimp in straw mulch) and avoiding compaction (minimize heavy equipment on seedbeds); (2) use sandy materials for seedbeds whenever possible, salvaging the sandiest topsoil or selectively handling sandstone overburden strata; and (3) take special care on steep south- and west-facing slopes, topdressing with the most favorable material or using a drought-adapted seedmix.

Geographic patterns.   In combination, the models in Fig. 5.7 indicate that different environmental factors are responsible for the variability in the success in revegetation at the different mines. Sand and bulk density are most important at Mines B and C, whereas sodicity and salinity dominate at Mines A and D. No factor emerges as a universal determinant of revegetation or stabilization at the four mines. (See Reith and Potter [1983] for more detailed interpretations of the models at each mine, and see Reith and Potter [1985] for more discussion of the geographic variation among the models in the San Juan Basin.)

The above interpretations have involved one factor at a time, because it is easiest to understand and explain the effect of a single determinant. However, multiple regression models consist of *combinations* of factors explaining variability in the outcome. The interpretors of multiple regression models should be aware of possible interactions among determinants, for instance, among slope and aspect or texture and sodicity.

Unexplained variability

The models in Fig. 5.7 explained up to 66.5% of the variability among the outcomes, although the high R^2 values at Mine D must be qualified by the relatively small sample size there. The average R^2 was about 40%, indicating that nearly 60% of the variability among the outcomes remains unexplained. Ecological modelers must not be discouraged by relatively low R^2 values, because natural processes are complex and not conducive to total analysis and explanation. As long as the significance level is below 0.05 (preferably *well* below), the modeler can be reasonably confident (1) that the relationships in the model portray real relationships in nature and (2) that factors which were tested, but which were not important in the model, are not highly correlated (at least not linearly) with the outcome.

The sources of unexplained variation in the case study are many, including uncontrolled and uncontrollable factors. In the former category are soil amendments, irrigation rates, topsoil depths, seed mixes, and other factors, which no doubt differed among the various sites at a mine, but which could not all be included in the multiple regression. These factors were excluded partly because the records of such treatments are vague and inaccessible, but mostly because there is an upper limit to the number of potential determinants which can be included in any analytical modeling effort. As more factors are included in the analysis, the sample size must be upwardly adjusted to an unachievable level.

Uncontrollable sources of variation are haphazard processes in nature that cannot be documented, no matter how intense the sampling effort. These include grazing and defecation by wildlife, microsite variation in climate and precipitation, and, most important, erratic seed application. The chaffy seeds of native grasses used for reclamation in the San Juan Basin frequently clog the apertures of the range drills which are used to press seed into seedbeds. The result is that some reclaimed surfaces, no matter how favorable the environmental conditions, are bare simply because they never received seed.

The explanatory powers (R^2) of regression models are reduced by these uncontrolled or uncontrollable sources of variation. However, the essential purpose of the statistical procedures which have been developed for analytical modeling is to identify the important relationships concealed within a complex network of natural events and processes.

Predictions with these models

The models in this case study were not intended for predictions, which is why the information required for predictions is not presented in Fig. 5.7. If it were desired, however, the models could be used to predict the biomass and frequency of perennial grasses, and the amount of erosion by rills, at sites on the mines where the necessary data were available. For instance, one could use the model for biomass at Mine C to predict biomass of perennial grasses (Y) based on measurements of bulk density ($X1$), percent sand ($X2$), age ($X3$), slope ($X4$), and percent stone ($X5$), simply by plugging the measured values into the following equation:

$$Y = -74.6X_1 - 133.1 (\text{Arcsine}[\sqrt{X_2}]) + 11.8X_3 - 2.2X_4 - 85.0 (\text{Arcsine}[\sqrt{X_5}]) + 9.10$$

The confidence in the prediction would be limited by the relatively low R^2 value. Furthermore, the effort would seem somewhat misplaced since biomass could actually be measured at the site. However, the model does enable rough predictions of biomass on future reclaimed surfaces based on estimations of the anticipated conditions for the five determinants in the biomass model. In this way, the model can be a valuable tool for management.

Future Applications of Modeling to Reclamation

Modeling has not been widely used in reclamation science, even though it is an accepted scientific technique that has been applied to numerous ecological phenomena. Possibly, some reclamation scientists resist modeling because it depends on approximation or simplification, and thus lacks the accuracy and confidence associated with straight-forward experimentation. However, the results of modeling are more general, more conducive to broad interpretation, than are the results of experimentation, which usually apply only to one specific set of circumstances. Furthermore, as emphasized from the outset, modeling organizes existing ideas and information into a more comprehensible format —into a context which invites continuing analysis and experimentation.

Ecosystem modeling

One fertile application of modeling in the future recognizes that the fundamental goal of reclamation is to restore natural function to a disturbed ecosystem. One of the first applications of large-scale computer modeling was toward analyzing and predicting the behavior of ecosystems (for example, the Grassland Biome Program; Innis 1978). The simulation model discussed earlier, by Ellis and Parton (1978), analyzed reclamation on the ecosystem level. The sophistication and accuracy of such modeling efforts should increase with the accumulation of knowledge and data pertaining to all aspects of ecosystems undergoing reclamation. In this regard, the modeler acts as synthesizer, continuously integrating new knowledge about reclamation into improved simulation models.

Predicting reclamation potential

Another promising application of modeling is toward the prediction of reclamation potential on as yet unmined lands. The analytical models by Packer et al. (1982) provide formulae for predictions of reclamation potential throughout the western United States. However, the appropriateness of the formulae and the accuracy of the predictions have not been fully tested. Use of the models should probably be confined to the Northern Great Plains states, from which baseline data were obtained.

However, the analytical modeling format in Packer et al. (1982) can serve as an excellent framework for the use of existing data and for the acquisition of new data to predict reclamation potential across relatively homogeneous geographic areas. The key to accuracy is to limit the scope of such efforts to physiographic areas which are ecologically bounded and which have a manageable degree of environmental variability.

The San Juan Basin, as defined and described in the opening chapter, is an example of an area over which predictive models should be reasonably accurate.

Predictive models of reclamation potential could accompany and quantify other assessment efforts (see Chapter 3) to give regional managers better foresight than they presently have on the consequences of their decisions to lease minerals and to approve mines. Reclamation potential was not quantified, and was therefore deemphasized, in the 1984 coal lease for the San Juan Basin, simply because there was no framework for the interpretation of the voluminous available data. Analytical modeling can enlighten future land management if integrated into the decison-making process.

Conceptual modeling: a good first step in all research

The broadest application of models to reclamation science should entail neither computers nor mathematics. Rather, conceptual models should be used at the inception of a project, regardless of the topic or method of study, to organize ideas and to provide a context for the research. Conceptual models can identify assumptions and facilitate interpretations. Model building can thus be used as an intellectual tool to improve the efficiency and insight of reclamation scientists.

References

Jeffers, J. N. R. 1978. An introduction to systems analysis: with ecological applications. London: Edward Arnold Ltd.

Ellis, J. E., and W. J. Parton. 1978. Impact of strip mine reclamation practices: a simulation model. Report to the Western Energy and Land Use Team, Office of Biological Services, U.S. Fish and Wildlife Service. Fort Collins, CO: Natural Resource Ecology Laboratory, Colorado State University.

Innis, G. S. 1978. Grassland simulation model. Ecological Studies no. 26. New York, NY: Springer Verlag.

Packer, P. E., C. E. Jenson, E. L. Nobel, and J. A. Marshall. 1982. Models to estimate revegetation potentials of land surface mined for coal in the West. U.S.D.A. Forest Service general technical report INT-123, Ogden, UT.

Parton, W. J., J. E. Ellis, and D. M. Swift. 1979. The impacts of strip mine reclamation practices: a simulation study. *In* M. K. Wali, ed., Ecology and coal resource development, Vol. 2, 584-591. New York, NY: Pergamon Press.

Reith, C., and L. D. Potter. 1983. An ecological analysis of vegetational reclamation at surface coal mines in New Mexico. Report to the Office of Surface Mining, Denver, CO.

Reith C., and L. D. Potter. 1985. Geographic variation among soil-based models of reclamation success. Journal of Arid Environments 8:223–234.

Rosenblueth, A., and N. Wiener. 1945. The role of models in science. Philosophy of Science 12.

Sokal, R. R., and F. J. Rohlfe. 1969. Biometry. San Francisco, CA: W.H. Freeman and Company, San Francisco, CA.

Wali, M. K. 1975. The problem of land reclamation viewed in a systems context. *In* M. K. Wali, ed., Practices and problems of land reclamation in western North America, 1-17. Grand Forks, ND: University of North Dakota Press.

6

The Selection of
Species for Revegetation

William W. Fuller

Introduction

Proper selection of species for revegetation is of major importance in successfully establishing vegetation in arid environments under dryland planting conditions, including those conditions which are encountered in the reclamation of mined lands. Method of seeding, depth of seed placement, seeding date, application of mulch, seeding rate, and other factors influence the ability to successfully establish vegetation. However, even if all factors are optimized, if the species selected are not suitable for the area, the planting is doomed to failure.

The purpose of this chapter is to discuss methods used in selecting species of plants for revegetation purposes and to review some of the species which are best adapted to arid and semiarid environments such as the San Juan Basin. Particular emphasis will be placed on how the Plant Materials Centers (PMCs), operated by the U.S. Department of Agriculture/Soil Conservation Service (USDA/SCS), select and release new varieties of plants. (Varieties are collections of plants with a distinct set of physical characteristics or properties which are retained when the plant reproduces.) The procedures used by PMCs are similar to procedures used by other agencies in the United States and elsewhere. The selection, testing, and release of species for commercial production is an ongoing task which produces a continually increasing array of plants for different environments with different conservation requirements.

Plant material centers

The primary purpose of the network of SCS Plant Materials Centers is to select superior plants for solving conservation problems. There are twenty-three of these centers in the United States. Plant Materials Centers are located so that they service several Major Land Resource Areas (MLRAs), which are delineated by similar climate, soils, and elevations, and which cross state lines (USDA/SCS 1981). Consequently, most PMCs serve two or more states. The center serving the San Juan Basin area is at Los Lunas, New Mexico, in the Rio Grande valley south of Albuquerque. The Los Lunas PMC serves all of New Mexico and parts of west Texas, eastern Arizona, southwestern Utah, and southeastern Colorado.

Collection and Evaluation of Plants

Accessions

The PMC usually obtains plants through a collection program organized by the State Plant Materials Specialist (PMS). This specialist, who is employed by the Soil Conservation Service, identifies local conservation needs and establishes priorities among species for collection.

Seed may also be obtained from sources abroad. A plant species which has only recently been brought to the United States from a foreign country is called an *exotic* species. Other species from overseas, called *naturalized* species, have long been established in the U.S., where they are spreading naturally. Both exotic and naturalized species are broadly referred to as *introduced* species.

Once it is decided which species will be collected, each PMS asks the county SCS field offices to find promising *accessions* of the species selected. An accession can be a single plant, seed from a single plant, or seed from a group of plants with similar characteristics. Each accession is given an identifying control number. Collectors find plants having some outstanding characteristic. Such plants might be producing an unusually large amount of seed, or be leafier than others in the area, or exhibit good growth characteristics in a very droughty site.

The location of the collection is carefully recorded to include legal description, MRLA, elevation, precipitation, topography, and soil type. Other plants growing in association are also recorded, along with pertinent remarks which explain the collectors' reasons for selecting this plant. Occasionally, rather than collecting seeds, a vegetative collection (all or part of a plant) is made. In any case, the collection of seeds or vegetative material is sent to the PMC, where it is transplanted or maintained and given a control number (the accession number).

At the end of the collection season, the PMC staff cleans the seed and gives it any treatments required to encourage germination. Some seeds from each accession are normally germinated in the greenhouse during the winter. At the proper time in the spring the young seedlings are transplanted to the field in what is known as an Initial Evaluation Planting.

Initial evaluation

The purpose of the initial evaluation is to observe and record the characteristics and comparative performance of all of the collected plant accessions of a single species as rapidly as possible so that promising plants can be selected for additional evaluation (USDA/SCS 1984). Included in this planting is the current *standard* for the species being evaluated, if one is available. The standard is the species or variety presently being used in the region for the intended conservation practice. If no standard is available, the performances of test species are compared against each other (USDA/SCS 1984). In any case, the plants are compared under identical growing conditions.

All of the plant accessions are allowed enough space and time to reach their full growth and potential. Rows from 15 feet to 25 feet long are commonly used for grasses, legumes, and forbs. The spacing of trees and shrubs is varied dependent upon the anticipated growth of the species and horticultural needs. Small plots may be used instead of rows to test species which spread by rhizomes or runners. Generally, fifteen herbaceous accessions or five woody accessions of a species are considered the minimum number for an adequate initial comparative evaluation. Extra rows of the standard are also planted on both sides of the tested plants to prevent having bare ground on one side of the tested row, which might bias the outcome.

Throughout this initial evaluation, the PMC specialists update a computer database with information on the ease of germinating the seed in the greenhouse, the percentage of germination, seedling vigor, and any other pertinent statistics and details.

Advanced evaluation

The next major step is to select those accessions which are equal or superior to the standard plant in one or more characteristics for the intended conservation use. In selecting superior plants, the PMC staff considers methods of propagation, fertility, and maintenance requirements. Plant Materials Center personnel, with assistance of the PMS and any other appropriate plant scientists who are available for consultation, base their selections of promising plants upon visual observations in combination with data from the computer.

After a plant is selected for advanced evaluation, the PMC specialists carefully screen all of the data for accuracy and establish a file for each selection. Then, they send a copy of the file to the regional SCS National Technical Center, where the regional PMS disseminates a description sheet to the national PMS and to colleagues and counterparts in his region and in other regions.

The advanced evaluation consists of intensive testing of plants that were superior during the initial evaluation. The objective is to determine quantitatively and qualitatively the potential for new plants to be better than the standards used for comparison. Cooperating agricultural experiment stations or other research agencies often become involved during this period. The advanced evaluation plants are frequently established in replicated rows or plots, and different treatment levels may be incorporated into the test based on the desires of the cooperating agency.

Special studies of establishment, management, and production are also conducted during advanced evaluation, as well as tests to determine whether the plant has any toxic qualities or potential for becoming a pest. If control is difficult, or if the plant has any toxic qualities, the situation is analyzed carefully before testing continues. Normally, plants with these potential problems are dropped from further consideration. Two to four accessions are usually selected as superior during the advanced evaluations, and seed-increase fields are planted at the PMC to provide enough seed for larger scale plantings for final evaluation.

Field evaluation plantings.   Field evaluation plantings (FEPs), not to be confused with field plantings (FPs), are established at locations in a different MLRA or on a site having soil, climate, or other conditions not represented at the PMC. These plot or row plantings are generally the responsibility of the PMC manager and are established by PMC personnel. The FEPs can be used for initial evaluations; however, they are more often used for advanced evaluations.

Final evaluations

Final evaluations of promising new plants are conducted in field plantings on private lands to assess their performance under actual use conditions. Field plantings are usually made on land owned or operated by farmers, by ranchers, or by mining companies, who have signed an agreement with the local Soil and Water Conservation District to cooperate in conservation activities on their land. Individuals who have such an agreement are called cooperators.

The tested plant materials are compared with the standard varieties under different soil, climate, and land use conditions. The field plantings are made on soils or sites where specific information is needed about the adaptation or comparative performance of a plant. These plantings are large enough to permit the use of conventional equipment and standard management methods. Detailed establishment and performance records are kept on the tested plants. Field plantings are also used to determine the geographic range over which the plant is adapted.

Planting guides. A planting guide prepared by the PMC includes a description of each plant, suggested potential uses, potential areas of adaptation, standards for comparison, soil adaptation, planting and maintenance procedures, and any other information about the plant that might have a bearing on its value and use. The PMC sends the guide to all states and field offices where the plant has a potential use.

When field office personnel locate cooperators who are interested in having a field planting on their farm, ranch, or mine, they contact the PMS. If the PMS approves the planting, seed is sent to the field office for the planting. The district conservationist, or a member of his staff, delivers the seed and ensures that the seedbed is properly prepared, that drills are properly calibrated, and that the general planting plan is followed.

Field planting evaluations

The evaluation results are presently tabulated by computer with sixty-six entries for grasses and forbs and seventy-nine entries for trees and shrubs. The performance of each plant is tested against one or more standards. Where there are no standards, the plant may be tested against another species used for the same conservation purpose, or it may be evaluated by itself.

Some of the major evaluation criteria include soil data; location; elevation; MLRA; seeding or planting information; climatic data; fertilizer rates; plant survival; seed production or spreading rate; resistance to insects, disease, salinity, and drought; grazing information; recovery ability following grazing; suitability for erosion control and wildlife habitat; forage clipping data, and recovery following clipping; irrigation information; and weed competition and control. Additional factors for woody plants are foliage density; attractiveness; height and width of crown; flowers; fruit; amount of litter and leaf drop; persistence of litter; value for windbreaks, screening, or noise abatement; and seasonal variation among these characteristics. During any evaluation, only some of the possible entries, as determined by the PMS, will actually be evaluated.

This phase of the testing requires close coordination between the SCS field office staff and the cooperator. The cooperator provides such information as amounts of fertilizer applied, amount and frequency of irrigation, stocking rates, length of time grazed, length of rest periods, and weed control data. Frequently, the field office staff will measure forage production by clipping and weighing the amount of growth produced during the growing season.

In New Mexico, field plantings are evaluated annually for the first five years, then placed in an inactive status for five years. Field plantings are evaluated during the inactive period only if some problem arises or if an unusual change occurs. A final evaluation is conducted during the tenth year following planting.

Release for commercial production

By the end of the first five years of the final testing phase, enough data has been accumulated from throughout the area of adaptation to ensure that the best accession can be selected. Also during this period, the PMC establishes larger seed-production fields to produce adequate seed for commercial seed producers. All of the seed is handled through the State Crop Improvement Association to ensure development of adequate safeguards for the maintenance of pure seed.

Examples of the selection process

Much of the work at PMCs concentrates on finding better varieties of species currently used for conservation. The result has been intermittent releases of new varieties with improved characteristics. Three examples are given below to illustrate that specific rules are not required for a release. Rather, the determination to release a variety is based upon a consensus among the plant scientists who are involved in the testing process.

Side-oats grama. In the species side-oats grama *(Bouteloua curtipendula)*, Vaughn side-oats was selected and released from the Los Lunas PMC in 1940 because of its adaptation to New Mexico conditions. However, Vaughn is a poor seed producer. El Reno, a variety of side-oats from Oklahoma, produces much more seed than Vaughn, although it is less well adapted to New Mexico. Seed producers raised El Reno because of its higher yields. Consequently, after several years on the market, Vaughn side-oats was generally in short supply. In recent years, in an attempt to produce a variety which is adapted to New Mexico, but which has high seed production, a new collection was made, resulting in the release of Niner side-oats in 1983. Niner was initially tested with twenty-eight other accessions, including both Vaughn and El Reno at the Los Lunas PMC. It was equal to the other accessions in seedling vigor, stand, seed production, and forage production. In a later planting, Niner was again equal to, or superior to, all accessions except for one which produced more forage and another which produced more seed. The high forage accession was a very poor seed producer and the high seed producer was a very low forage producer in this test (Quinones 1972, 1980).

In thirty-six field tests in New Mexico and Colorado, Niner had the best stand and vigor in twelve plantings and Vaughn was best in fourteen. They were equal in ten. However, during a 16-year trial period at the Los Lunas PMC, Niner averaged 50 pounds more pure live seed per acre than Vaughn. During these sixteen years, Vaughn exceeded Niner only twice, once in 1969 by 9 pounds and again in 1972 by 42 pounds. Niner exceeded Vaughn in the other fourteen years, once by as much as 327 pounds.

Blue grama.   In the case of blue grama *(Bouteloua gracilis)*, an initial evaluation of twenty-one accessions was planted at the Los Lunas PMC in 1958. From this IEP, a variety named Lovington was released in 1963 due to its high forage production. The initial evaluation planting was established under irrigation and was irrigated through 1960. The IEP planting was abandoned in 1961 but not destroyed; no irrigation was applied through 1964. Rainfall during the growing seasons for 1961 through 1964 was 5.7, 4.8, 4.9, and 3.6 inches, respectively. At the end of the 1964 growing season, it was noticed that eighteen of the accessions had died, two accessions had only traces of plants left, while an accession which was eventually named Hachita had maintained a 30% stand. Seed from these remaining plants of Hachita was increased for further testing.

Tests by A. M. Wilson, of the U.S. Department of Agriculture, Agricultural Research Service, suggested a reason for the drought tolerance of Hachita. Wilson showed that the number of roots per seedling, the rate of root growth, the root density, and the total weight of roots per seedling were all significantly higher than for the other accessions tested (USDA/SCS 1980). Quinones (1974) found that Hachita did not appear to be significantly different from Lovington in the characteristics he studied, except that Hachita appeared to produce more seed under unfavorable conditions, which may have been due to the better root system described by Wilson.

Seventeen field evaluation plantings were established in an area which extended from southwestern Utah and central Colorado in the north to southwestern New Mexico in the south. Hachita exceeded Lovington in establishment and performance in eleven plantings; the two were equal in four plantings; and Lovington exceeded Hachita in two plantings (USDA/SCS 1980). Hachita was obviously more drought tolerant than any blue grama currently on the market, and although the forage yield was not significantly better than that of Lovington during the tests, the data indicated that Hachita yields improved annually for the first several years. Based on the above information and other data, Hachita was named and released in 1980.

Introduced yellow bluestem. Introduced plants normally go through a much more vigorous field testing, since they must exhibit characteristics which are better than the plants native to this country. For example, Ganada yellow bluestem *(Bothriochloa ischaemum)* was first introduced from Turkestan in 1934, but was not released until 1979. During this period it went through the normal initial and advanced evaluations and then moved into the field testing program. In field tests, Ganada was planted in thirty-three trials all across New Mexico, in eighteen locations in Colorado, and in twenty-two locations in Arizona. All of these were dryland plantings. In one of the New Mexico tests, Ganada still had an excellent stand of grass after nineteen years, which gave it the best rating of any of the twenty-two native species with which it was compared. In a southwestern Colorado test, Ganada and only one of the twenty-two native species, a spike muhly, were excellent after sixteen years of comparison. All of the other species received lower ratings.

Ganada had proved to be well adapted to the area. It did well in northeastern Colorado, just south of Colorado's northern border, at an elevation of 5,300 feet and with 11 inches of annual precipitation, and in central Colorado (6,000 ft. elevation and 14 inches ppt.). At the southern extreme, Carlton Herbel, Director of the Agricultural Research Service's Jornada Experimental Range at Las Cruces, New Mexico, found that after clearing creosote bush, Ganada was the only grass that could be regularly established on the clays and clay loams (Herbel 1980). This area has an elevation of 3,880 feet and receives 9 inches of annual precipitation. Ganada also did well in the twenty-two Arizona plantings. Based on its ability to maintain excellent stands under such a wide variation in climate, Ganada was named and released. It has recently been tested for forage production and adaptability at the Noble Foundation, located in southern Oklahoma near Ardmore. After three years of testing, Ganada produced over 500 pounds more oven dry forage than Plains yellow bluestem (also an introduced variety), the standard for comparison in that area (Dalrymple et al. 1982). Presently, Ganada's full area of adaptation is unknown.

Wildlife considerations

The establishment of wildlife habitat is generally a secondary consideration in most revegetation efforts, at least for relatively small-scale projects such as mine reclamation. Many agronomists believe that wildlife value is a more or less automatic consequence of successful revegetation. The most important ingredient for wildlife habitat is *structural diversity* in the plant community, meaning that the vegetation has a mix of growth forms: trees, shrubs, grasses, and forbs. In research under way at mines in Wyoming and Montana, Stoecker (1984) has found plant structural diversity to be far more important than species diversity in determining how many individuals and species of wildlife visit reclaimed surfaces (personal communication). Thus, long-term reclamation plans should include the establishment of some woody species along with the perennial grasses used for stabilization and grazing by livestock. A woody planting can be especially attractive to wildlife if the selected species are among those known to be desirable for browsing.

Introduced versus native species for revegetation

The use of introduced species for revegetation in many parts of the West has caused much controversy. Many believe that only native species should be used. According to Thornburg (1982), there is some justification for the opposition to introduced species, because many are shorter lived than native species in arid and semiarid areas. However, Thornburg also points out that many exotic species have become naturalized and perform better than native species. An excellent example is Ganada yellow bluestem, which has been widely tested and has provided excellent stands of grass in many cases where nearly all of the native plants failed.

Guidelines for plant selection

Much of the discussion so far is summarized by a model by McKell et al. (1982) for the identification of plants for revegetating a specific site. The model consists of a hierarchy of criteria by which available plant species are evaluated for their suitability for the site. The criteria consider site characteristics which are evaluated by sampling the environment, including the native vegetation. Those plant species which best conform to the site characteristics are most suitable for use. The criteria for the model are presented in Table 6.1

The foremost consideration in the selection of species for revegetation is the purpose that the vegetation is meant to accomplish after it is established. A planting made specifically for erosion control could be made up of entirely different species from a planting which is to be used primarily for livestock grazing.

A good example is the use of 'Sodar' streambank wheatgrass *(Agropyron riparium)*. This is a native sod-forming grass which could be used

Table 6.1.   A hierarchy of criteria specified by McKell et al. (1982) for selecting plants to be used in the revegetation of a specific site. In this example, the post-mining land use is grazing.

1. Capability for propagation and establishment under local stresses.
   A. Ease of obtaining seed or other vegetative material.
   B. Ease of propagation.
   C. Ease of planting.
   D. Immediacy and certainty of establishment.

2. Value for animal nutrition.
   A. Forage value.
      Palatability
      Nutritive value
      Tolerance to grazing
      Stability or dependability of production
      Length of green period
      Accessibility of forage
   B. Growth rate.
      Aboveground growth rate and forage production
      Below-ground growth rate
      Cover
      Season of maximum growth

3. Adaptability to climatic extremes.
   A. Tolerance to drought.
   B. Tolerance to temperature extremes.
   C. Tolerance to strong winds.

4. Adaptability to soil conditions.
   A. Soil-water relations.
   B. Tolerance to salinity.
   C. Tolerance to unfavorable pH.
   D. Tolerance to nutrient deficiencies and toxicities
   E. Tolerance to wetness and waterlogging.

5. Protection of watershed (stabilization).
   A. Below-ground growth rate.
   B. Rate of spreading and vegetative cover.

6. Suitability for miscellaneous conditions.
   A. Persistence.
   B. Self-renewal.
   C. Compatibility with other species.
   D. Disease and pest resistance.
   E. Fire resistance.
   F. Value for landscaping and other aesthetic purposes.
   G. Minimum maintenance costs.

in the San Juan Basin. It is a good erosion-control species due to its drought tolerance and its sod-forming abilities. However, it has tough leaves and stems because of its very high cellulose and lignin content, and it is seldom grazed by livestock. Consequently, a high percentage of Sodar would be undesirable for a planting for which grazing is the primary purpose, because livestock gains would be poor. On the other hand, where both erosion control and some livestock forage production is desired, one might mix a small amount of Sodar seed with that of more palatable species to ensure a good ground cover. A mix such as this would be particularly helpful in areas where overgrazing is a problem.

The vegetation in natural plant communities located close to a disturbed area can guide a plant-selection program. However, one should remember that the rangeland vegetation in the Basin has changed greatly since the area was settled in the late 1500s. Most of the areas have deteriorated to a brush-dominated subclimax community. Many grass and forb species that would be found in a climax situation are either absent or are present in very low numbers. Consequently, attempts should not be made to reproduce the present natural communities. Rather, these communities should be used only as a guide for plants which could be included in the planting.

According to Oaks (1982), there are more than four thousand native or naturalized plant species or subspecies which occur in the Southwest. The species selected for reseeding a specific site should be suited to the intended use and management, and be well adapted to the soil, climate, and topography. This selection requires a blend of ecological and agronomic principles.

Plant Selection for Surface Mines in the San Juan Basin

The San Juan Basin presents a difficult challenge to the plant scientist intent on revegetating disturbed areas. Average annual precipitation in the area varies from about 6 inches to 15 inches. Much of the rainfall in the 7-9 inch zone comes during the summer months in short-duration, high intensity thunderstorms (U.S. Department of Commerce 1941). Yearly precipitation varies widely from the average. Temperatures in the Basin are extreme, ranging from a high of 110° F at Fruitland to a low of -30° F at Bloomfield. The growing season in the San Juan Basin varies from about 145 days at the southern end to over 160 days in the central part. The frost-free period is generally from early May to late October (U.S. Department of Commerce 1964). Due to the high summer temperatures and scant cloudcover during the rainy season, evaporation in the region is high. The water deficit for the region is from 15 to 17 inches, resulting in most of the region being classified as arid or semiarid. The semiarid portion of the region differs from the arid part by having slightly lower temperatures and slightly greater annual precipitation (State Planning Office 1973).

Species and varieties for the Basin

The San Juan Basin is served by the Los Lunas PMC, which has been in existence for forty-four years. In cooperation with other agencies, the PMC has released thirty improved varieties of plants during this period, including twenty grasses, one forb, and nine woody species. All of these plants are well adapted in New Mexico and have been superior in nearly all respects to plants from native seed sources. Table 6.2 lists New Mexico releases for the San Juan Basin, including their intended conservation use.

The New Mexico state government, universities, and local mining companies have all sponsored reclamation programs for mined land and other disturbed areas. One of the earliest revegetation programs in the Basin was motivated by the New Mexico State Highway Department. In 1965, the Department contracted with the SCS to assist them not only in selecting species for revegetation, but also in developing methods of revegetating roadsides. This project covered the entire state and ended in 1974. Although work was done in the San Juan Basin, general recommendations were made for the entire state rather than specifically for the Basin area. In 1982, the Highway Department again contracted with the SCS for work in species selection and establishment procedures. Presently, all of the recent roadside seedings in the San Juan Basin are being evaluated annually as to the success of different species and specific methods of establishment. Several revegetated borrow pits in the area are also being monitored.

Research projects at older surface mines in the Basin

The surface mining industry also has a great interest in revegetating disturbed areas. At present, twelve mines operate in the San Juan Basin, all of which are legally bound to revegetate any areas they mine (Reith and Potter 1983, Roybal and Eveleth 1982). Several of these mines have contracted with plant scientists for assistance in the selection of species and the development of establishment methods.

Mine B.   Mine B contracted with Howard Stutz, Professor of Botany and Genetics, Brigham Young University, to select and breed adapted shrubs. Stutz (1982a) established three sets of experimental plots at the mine: 1) selection plots, 2) testing plots, and 3) seed increase-selection plots. The selection plots were used for initial identification of superior candidates for the breeding program. Chenopod shrubs dominated the mine area; consequently, they made up the majority of the plants tested. Although most of the introductions failed, Stutz found enough successful species to provide adequate material for further consideration.

As superior performers became conspicuous in the selection plots, they were advanced to the testing plots. Here they were tested for their response to fertilizer and water, as well as to the forces of natural selection. During the testing, Stutz eliminated many accessions and ended up with five saltbush *(Atriplex)* hybrids. These hybrids were moved to one-acre seed increase-selection plots, where seed is produced for large-scale application to mine spoil. In these plots, Stutz looked for plant

Table 6.2. A chronology of plant releases by the New Mexico Plant Materials Center (Los Lunas).

Date	Variety	Species	Conservation use
1940	Vaughn	sideoats grama	Erosion control and range seeding
1961	*Largo*	tall wheatgrass	Pasture on saline-sodic soils
1963	Lovington	blue grama	Erosion control and range seeding
1963	*Luna*	pubescent wheatgrass	Erosion control and forage
1965	*Jose*	tall wheatgrass	Pasture on saline-sodic soils
1973	Arriba	western wheatgrass	Erosion control and forage
1973	Redondo	Arizona fescue	Erosion control and range seeding
1973	El Vado	spike muhly	Erosion control and range seeding
1973	*Pink lady*	winterberry euonymus	Windbreaks, beautification, and wildlife
1973	Bandera	Rocky Mtn. penstemon	Erosion control and beautification
1974	Paloma	Indian ricegrass	Erosion control and range seeding
1978	Montane	mountain-mahogany	Wildlife
1978	Jemez	New Mex. forestieria	Erosion control, windbreaks, and wildlife
1978	*King red*	Russian olive	Wildlife and windbreaks
1979	Bighorn	skunkbush sumac	Erosion control and wildlife
1979	Viva	galleta	Range seeding and mine reclamation
1979	*Ganada*	yellow bluestem	Range seeding and roadside erosion control
1980	Hachita	blue grama	Range seeding and mine reclamation
1982	Salado	alkali sacaton	Range seeding on saline-sodic soils and mine reclamation
1982	Autumn amber	skunkbush	Erosion control and beautification
1983	Niner	sideoats grama	Range seeding and erosion control

*All varieties are adapted for dryland revegetation in the San Juan Basin. Italicized varieties are introduced. Scientific names are in Table 6.3.

response to natural selection and for such attributes as desirable growth form, adequate seed production, high seed germination, high capacity for seedling establishment, and capacity to respond favorably to machine harvesting, which was one of the major selection factors. After machine harvesting, only those plants which produced another seed crop were included in the selected product. Seed from a number of such plants were then used to plant another one-acre plot in which machine harvesting was again used. A third round of selection and machine harvesting followed, and the process was continued until a satisfactory product was obtained (Stutz 1982a).

In each round of selection, Stutz deliberately included a large number of top performers with the rationale that plants selected for use on disturbed lands must have a wide genetic base in order to tolerate the environmental variability on reclaimed surfaces which are undergoing natural succession. (Stutz 1982b).

Mine A.   Quinones (1980) was funded by the U.S. Department of Agriculture and the Office of Research and Development of the U.S. Environmental Protection agency to evaluate species at Mine A. Between 1976 and 1980, he studied plants collected and cataloged from several academic and government institutions. For each species, Quinones conducted two to four plantings, each of which was irrigated for two growing seasons following seeding (see Chapter 7). The third-year planting consisted of the best performers for the first two years, and the fourth-year planting consisted of the best performers from the three previous years. In all, Quinones evaluated two hundred and sixty-three accessions of twenty-three species of grass, eight accessions of four forbs, and twenty-seven accessions of eight species of shrubs.

The results were disappointing. Only one grass, alkali sacaton *(Sporobolus airoides)*, did well throughout the test period. Two other grasses, galleta *(Hilaria jamesii)* and blue grama, did well from 1975 through 1978, but deteriorated thereafter, and the shrub four-wing saltbush *(Atriplex canescens)* did well during 1977 and 1978, but then deteriorated. The failure of so many species and varieties was initially perplexing; however, subsequent analyses revealed potentially inhibitory levels of sodium and other salts in the spoil underlying the 15-cm deep mantle of topsoil. The plants grew well in the topsoil, but died when their roots reached the saline/sodic conditions of the shaley spoil. Irrigation may postpone this process by temporarily leaching salts beyond the root zone, but they will likely migrate upwardly when supplemental water is withdrawn. Alkali sacaton probably survived because it is more tolerant of salts in the soil than most other grasses (Burnstein 1958).

Mine A has since contracted with the SCS to conduct another investigation of species and seeding methods for use in its reclamation effort. In this study, thirty-two accessions of eleven species of grass, one accession each of four species of forbs, and eight accessions of three species of shrubs are being evaluated. The SCS made plantings in 1981, 1982, and 1983 which are irrigated for only the first two years, allowing initial establishment and subsequent evaluation of survival. Since the study is just beginning, no conclusions have been drawn (USDA/SCS 1982).

Revegetation research at a recent mine in the Basin

In 1981, Soil Conservation Service plant specialists began a similar study for a coal company at a recently developed mine site in the center of the San Juan Basin, approximately thirty miles SSE of the vicinity of Mines A and B. This small surface mine is similar in geology and climate to Mines A and B, but lacks the relatively high quality irrigation water available at the other two mines. Instead, irrigation water was used that contained high levels of sodium, resulting in the formation of water-repellent surface crusts and, to some extent, inhibition of germination. The plantings were irrigated with this sodic water for only the first two years. In the first year, 16 inches of water were applied, mostly in one-inch applications at gradually increasing intervals. In the second year, one inch of water was added every two weeks for eight weeks. There are sixteen accessions of twelve species of grass, one accession each of three species of forbs, and three accessions of two species of shrubs undergoing evaluation in this continuing study.

The preliminary data has been confounded by the poor water quality. The researchers are not sure whether poor plant performance is due to a lack of adaptation to the soils and climate or due to the poor water (USDA/SCS 1982).

A list of species for revegetation in the San Juan Basin

The most comprehensive review of species for revegetation of surface mines in arid and semiarid lands in the United States has been assembled by Thornburg (1982). He lists two hundred or more species of grasses, forbs, and woody plants which are adapted to the arid and semiarid West. Although Thornburg lists one hundred and ten species as being adapted to the San Juan Basin, we regard seventy-six of these as especially appropriate. These include twenty-six species of grasses (nine exotic), six species of forbs (four exotic), and forty-one species of woody plants (three exotic). Table 6.3, which lists these species and some of their performance characteristics, is adapted from Thornburg's review of plant materials for the western U.S.

Table 6.3. Specific characteristics and adaptations to moisture, soil, and adapted varieties. *see note on page 130.

[1]B—bunch; S—sodformer; D—deciduous; E—evergeen.

[2]W—warm season; C—cool season; A—annual; B—biennial; P—perennial; T—tree; S—shrub

Plant performance: ■ Optimum   ▨ Below optimum

Species:	N—native, I—introduced	B or S; D or E[1]	W or C; A, B, or P; T or S[2]	Precipitation zone (Inches) 8 12 16 20	Adaptation to soils (Sand, Sandy, Silty, Clayey, Clay)	Shallow sites	Calcareous sites	Salt tolerant	Acid tolerant	Adapted varieties
Grasses										
Agropyron cristatum (crested wheatgrass)	I	B	C							Ephraim, Ruff, Fairway
Agropyron dasystachyum (thickspike wheatgrass)	N	S	C							Critana
Agropyron desertorum (standard crested wheatgrass)	I	B	C							Nordan
Agropyron elongatum (tall wheatgrass)	I	B	C						X		Jose, Largo
Agropyron riparium (streambank wheatgrass)	N	S	C					X		Sedar
Agropyron sibiricum (Siberian wheatgrass)	I	B	C							
Agropyron smithii (western wheatgrass)	N	S	C					X		Arriba
Agropyron trachycaulum (slender wheatgrass)	N	B	C					X		San Luis, Primar

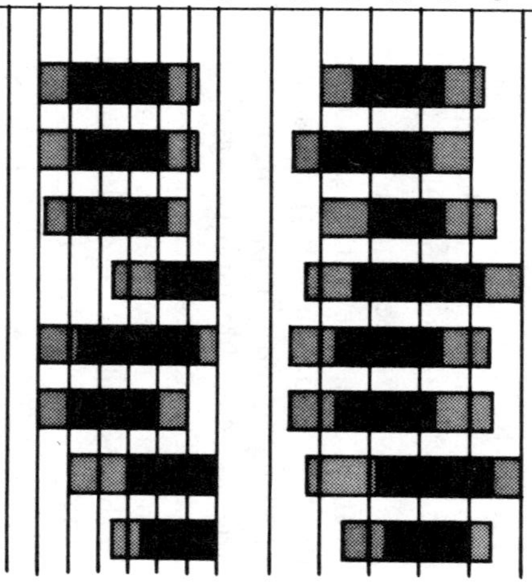

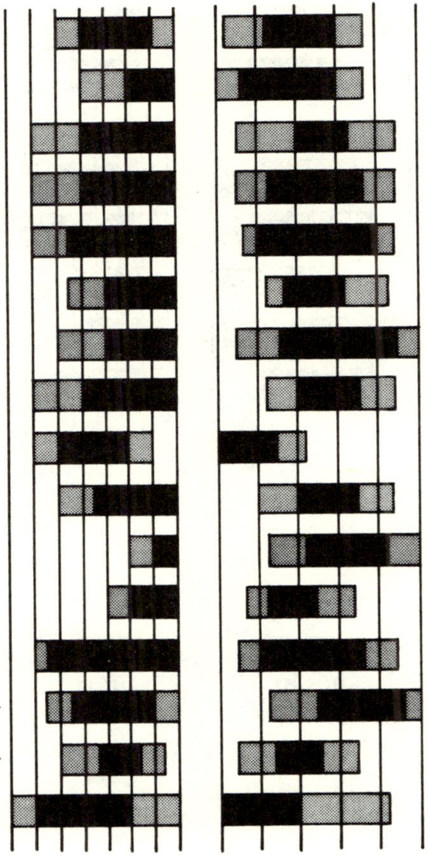

Species								Cultivar
Agropyron trichophorum (pubescent wheatgrass)	I	S	C					Luna
Andropogon scoparius (little bluestem)	N	B	W	X	X			Pastura
Bothriochloa ischaemum (yellow bluestem)	I	B	W	X	X			Ganada
Bouteloua curtipendula (sideoats grama)	N	B	W	X	X			Niner, Vaughn
Bouteloua gracilis (blue grama)	N	S	W					Hachita, Lovington
Dactylis glomerata (orchardgrass)	I	B	C				X	Berber, Paiute
Distichlis stricta (inland saltgrass)	N	S	W			X		
Elymus cinereus (basin wildrye)	N	B	C			X		Magnar
Elymus giganteus (mammoth wildrye)	I	S	C			X		Volga
Elymus junceus (Russian wildrye)	I	B	C			X		Vinall
Elymus triticoides (beardless wildrye)	N	S	C			X		
Festuca arizonica (Arizona fescue)	N	B	C	X				Redondo
Festuca ovina (sheep fescue)	N	B	C					Covar
Hilaria jamesii (galleta)	N	S	W					Viva
Muhlenbergia wrightii (spike muhly)	N	B	W					
Oryzopsis hymenoides (Indian ricegrass)	N	B	C					Paloma, Nezpar

(continued on following page)

125

Table 6.3. *see note on page 130.*

[1]B—bunch; S—sodformer; D—deciduous; E—evergeen.

[2]W—warm season; C—cool season; A—annual; B—biennial; P—perennial; T—tree; S—shrub

Plant performance: ■ Optimum   ▨ Below optimum

Species:	N—native, I—introduced	B or S; D or E[1]	W or C; A, B, or P; T or S[2]	Precipitation zone (Inches: 8 12 16 20)	Adaptation to soils (Sand, Sandy, Silty, Clayey, Clay)	Shallow sites	Calcareous sites	Salt tolerant	Acid tolerant	Adapted varieties
Grasses										
Poa canbyi (Canby bluegrass)	N	B	C			X				Canbar
Sporobolus airoides (alkali sacaton)	N	B	C					X		Salado, Saltalk
Sporobolus contractus (spike dropseed)	N	B	W			X				
Sporobolus cryptandrus (sand dropseed)	N	B	W			X	X			
Stipa comata (needle-and-thread)	N	B	C			X	X			
Forbs										
Erodium cicutarium (alfilaria)	I	B	A							
Kochia prostrata (prostrate summer cypress)	I	B	P			X		X		Immigrant

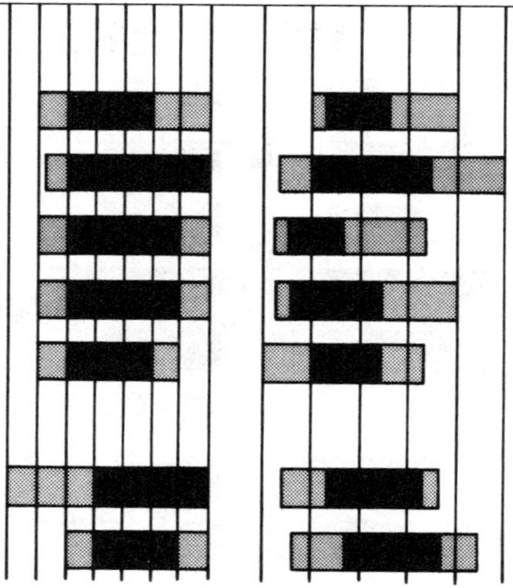

126

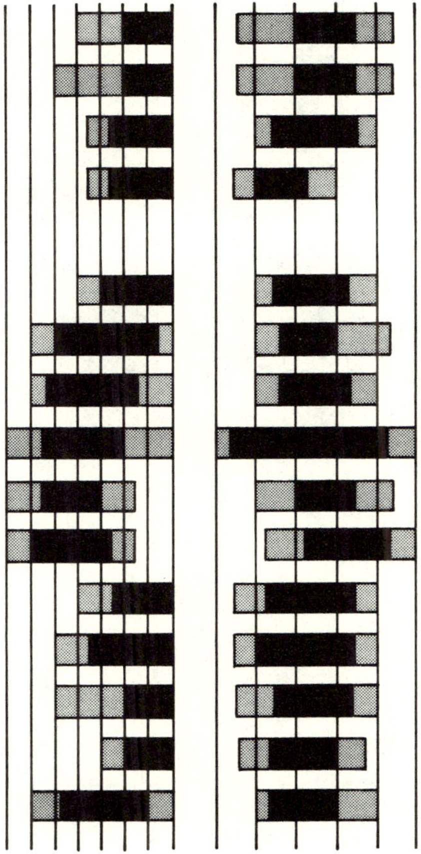

Species				Notes
Melilotus alba (white sweet clover)	I	B	B	
Melilotus officinalis (yellow sweet clover)	I	B	B	
Penstemon palmeri (Palmer penstemon)	N	S	P	
Penstemon strictus (Rocky Mountain penstemon)	N	B	P	Bandera
Woody plants				
Amelanchier spp. (serviceberries)	N	D	S	
Artemisia arbuscula nova (black sagebrush)	N	D	S	
Artemisia tridentata (big sagebrush)	N	D	S	
Atriplex canescens (fourwing saltbush)	N	E	S	X X Rincon, Wytana
Atriplex lentiformis (quailbush)	N	E	S	X
Atriplex nuttallii (Nuttall saltbush)	N	E	S	X
Ceanothus fendleri (Fendler ceanothus)	N	D	S	
Ceanothus greggii (desert ceanothus)	N	E	S	
Ceanothus martinii (Martin ceanothus)	N	E	S	
Ceanothus prostratus (squawcarpet)	N	E	S	
Cercocarpus ledifolius (curl-leaf mountain-mahogany)	N	E	S	X X

(continued on following page)

127

Table 6.3. *see note on page 130.*

	Plant performance						
				■ Optimum	Below optimum ▨		Adapted varieties
[1]B—bunch; S—sodformer; D—deciduous; E—evergeen.				Precipitation zone	Adaptation to soils	Sites	
[2]W—warm season; C—cool season; A—annual; B—biennial; P—perennial; T—tree; S—shrub	N—native, I—introduced	B or S; D or E[1]	W or C; A, B, or P; T or S[2]	Inches 8 12 16 20	Sand Sandy Silty Clayey Clay	Shallow sites / Calcareous sites / Salt tolerant / Acid tolerant	Adapted varieties
Species:							
Woody plants							
Cercocarpus montanus (true mountain-mahogany)	N	D	S			X X (Shallow, Calcareous)	Montane
Chrysothamnus nauseosus (rubber rabbitbrush)	N	D	S			X X (Calcareous, Salt tolerant)	
Cowania mexicana (cliffrose)	N	E	S			X X (Shallow, Calcareous)	
Elaeagnus angustifolia (Russian olive)	I	D	T				King Red
Ephedra viridis (green ephedra)	N	D	S				
Eriogonum umbellatum (sulfur eriogonum)	N	D	S				
Euonymus bungeonus (winterberry euonymus)	I	D	S				
Eurotia lanata (winterfat)	N	E	S			X X X (Shallow, Calcareous, Salt tolerant)	

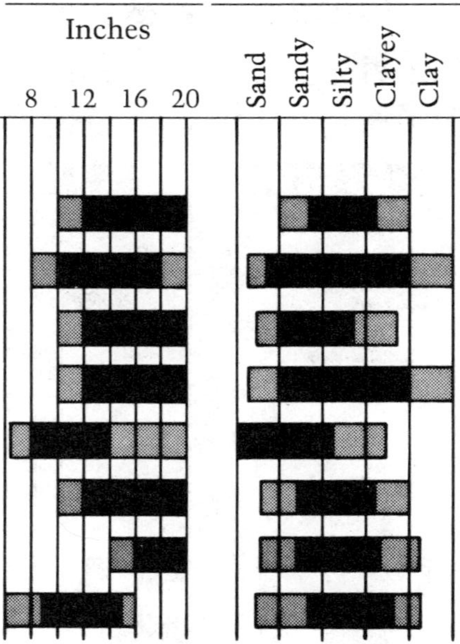

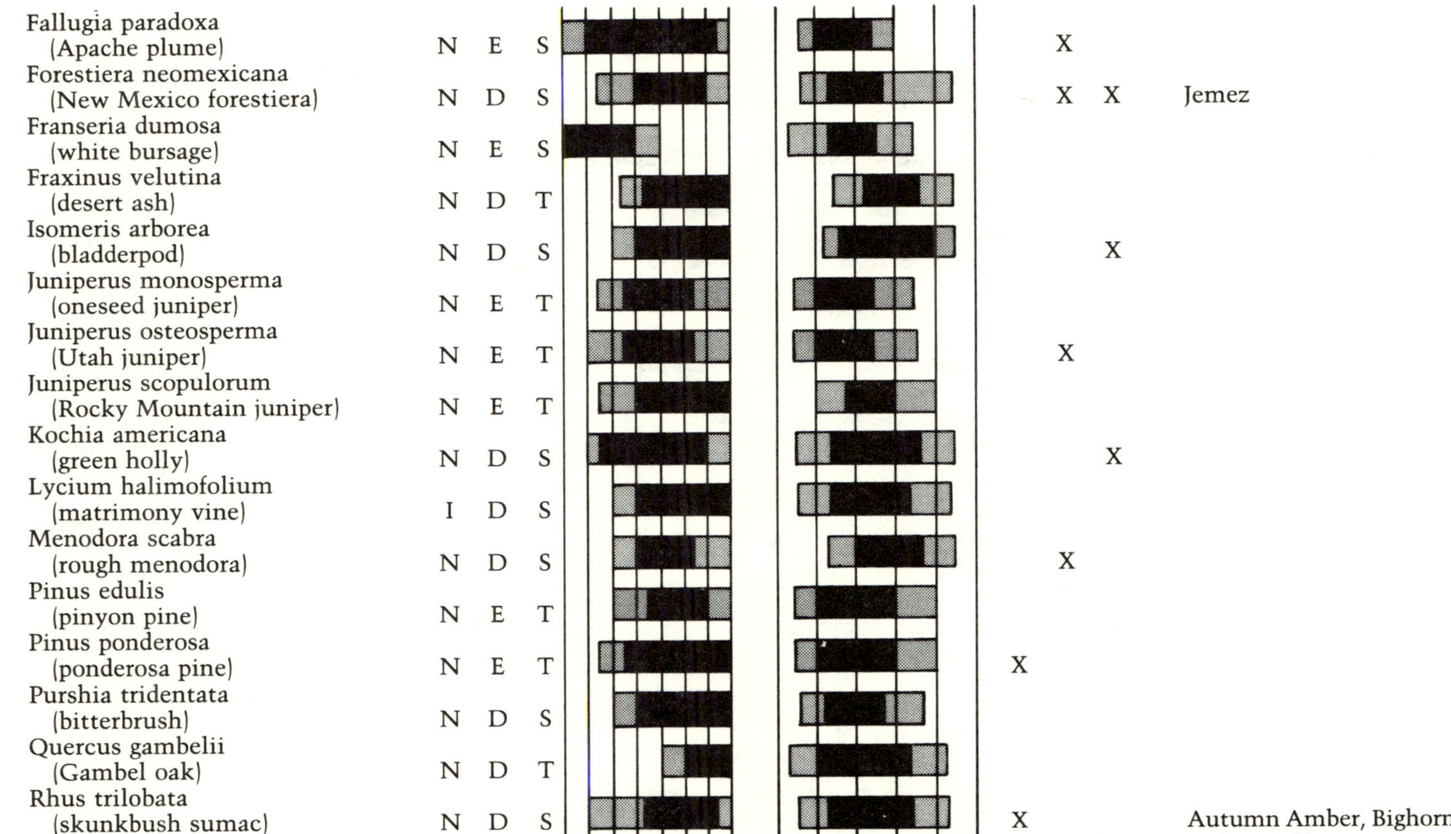

Species				Variety
Fallugia paradoxa (Apache plume)	N	E	S	X
Forestiera neomexicana (New Mexico forestiera)	N	D	S	X X Jemez
Franseria dumosa (white bursage)	N	E	S	
Fraxinus velutina (desert ash)	N	D	T	
Isomeris arborea (bladderpod)	N	D	S	X
Juniperus monosperma (oneseed juniper)	N	E	T	
Juniperus osteosperma (Utah juniper)	N	E	T	X
Juniperus scopulorum (Rocky Mountain juniper)	N	E	T	
Kochia americana (green holly)	N	D	S	X
Lycium halimofolium (matrimony vine)	I	D	S	
Menodora scabra (rough menodora)	N	D	S	X
Pinus edulis (pinyon pine)	N	E	T	
Pinus ponderosa (ponderosa pine)	N	E	T	X
Purshia tridentata (bitterbrush)	N	D	S	
Quercus gambelii (Gambel oak)	N	D	T	
Rhus trilobata (skunkbush sumac)	N	D	S	X Autumn Amber, Bighorn

(continued on following page)

Table 6.3.

Species:	N—native, I—introduced	B or S; D or E[1]	W or C; A, B, or P; T or S[2]	Precipitation zone (Inches: 8, 12, 16, 20)	Adaptation to soils (Sand, Sandy, Silty, Clayey, Clay)	Sites: Shallow sites	Sites: Calcareous sites	Sites: Salt tolerant	Sites: Acid tolerant	Adapted varieties
Woody plants										
Ribes aureum (golden currant)	N	D	S							
Robinia neomexicana (New Mexico locust)	N	D	T							
Salix spp. (willows)	N	D	S							
Shepherdia argentea (silver buffaloberry)	N	D	S					X		
Symphoricarpos albus (common snowberry)	N	D	S							
Tamarix aphylla (athel tamarisk)	I	E	T					X		

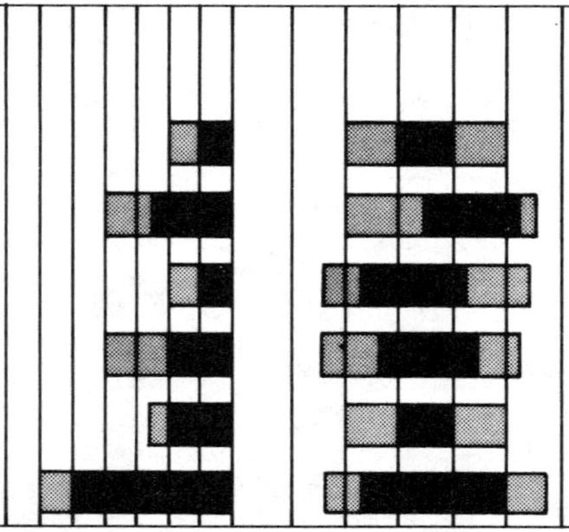

Plant performance: ■ Optimum ▒ Below optimum

[1]B—bunch; S—sodformer; D—deciduous; E—evergeen.

[2]W—warm season; C—cool season; A—annual; B—biennial; P—perennial; T—tree; S—shrub

*Specific adaptation and performance may vary by localized area and intended use. "Below optimum" indicates that the selection may be outperformed by other species or is not competitive.

Many of the species in Table 6.3 do not have commercially developed varieties, as indicated wherever the variety column is blank. Materials for these plants must be collected by wild harvests. Where more than one commercial variety is available, local dealers should be consulted to determine the best variety for a specific site.

Summary

Species selection should result from a cooperative effort between government agencies and private concerns. The Soil Conservation Service guides this effort by operating Plant Materials Centers which provide expertise and other resources for selecting, testing, and releasing species. Decisions are based on a complex mixture of scientific principles, experimental results, and conservation needs. The examples herein demonstrate the variety of methods for selection and provide guidelines for future work. The review of species adapted for the San Juan Basin illustrates the number and variety of species which can be made available for revegetation in regions with comparable environmental variability.

However, one should realize that the process of species selection is never-ending, because new conservation needs will always arise, and improved species or varieties will be selected or developed to address these needs.

References

Bernstein, L. 1958. Salt tolerance of grasses and forage legumes. U.S. Department of Agriculture, Agricultural Research Service, Agricultural information bulletin no. 194.

Dalrymple, R. L., J. Rogers, and L. Timberlake. 1982. Asiatic bluestems. Agriculture Division, Noble Foundation report AB-1982.

Hassell, W. G. 1982. New plant materials for reclamation. *In* E. F. Aldon and W. R. Oaks, eds., Reclamation of mined lands in the Southwest, 108-112. Albuquerque, NM: New Mexico chapter, Soil Conservation Service Society of America.

Herbel, C. 1980. Director, ARS Jornada Experimental Range, Las Cruces, NM. Personal communication.

Marquiss, R. W., L. E. Bartel, and G. G. Davis. 1974. Improved forage species for reseeding in the San Juan Basin. Colorado State University Experiment Station technical bulletin 122.

McKell, C. M., J. Briede, and S. Pendleton. 1982. Selection of plant materials for the Southwest. Paper presented at the Reclamation of Mined Lands in the Southwest Symposium, Albuquerque, NM, October 20-22.

Oaks, W. R. 1982 Reclamation and seeding of plant materials for reclamation. *In* E. F. Aldon and W. R. Oaks, eds., Reclamation of mined lands in the

Southwest, 145-150. Albuquerque, NM: New Mexico chapter, Soil Conservation Service Society of America.

Quinones, F. A. 1972. Evaluation of range grasses under irrigation. New Mexico State University Agricultural Experiment Station research report 224.

Quinones, F. A. 1974. Performance of blue grama, sideoats grama, and alkali sacaton accessions under irrigation near Las Cruces, New Mexico. New Mexico State University Agricultural Experiment Station research report 289.

Quinones, F. A. 1980. Evaluation of range plants at the San Juan Mine, near Farmington. New Mexico State University Agricultural Experiment Station research report 289.

Reith, C., and L. D. Potter. 1983. An ecological analysis of vegetational reclamation at five surface coal mines in New Mexico. Report to the Office of Surface Mining, Denver, CO.

Roybal, G., and R. W. Eveleth. 1982. Overview of surface minining in New Mexico and ongoing reclamation projects. *In* E. F. Aldon and W. R. Oaks, eds., Reclamation of mined lands in the Southwest, 2-7. Albuquerque, NM: New Mexico chapter, Soil Conservation Service Society of America.

State Planning Office. 1973. The climate of New Mexico. Revised edition.

Stoecker, R. E. 1984. Wildlife and vegetation diversity. Stoecker-Keammerer and Associates, Ecological Consultants, Boulder, CO. Personal communication.

Stutz, H. C. 1982a. Breeding superior plants for disturbed lands. Paper presented at the Western Mined Land Rehabilitation Research Workshop, U.S. Forest Service, Fort Collins, CO, June 10-11.

Stutz, H. C. 1982b. Broad gene pools required for disturbed lands. *In* E. F. Aldon and W. R. Oaks, eds., Reclamation of mined lands in the Southwest, 113-118. Albuquerque, NM: New Mexico chapter, Soil Conservation Service Society of America.

Thornberg, A. A. 1982. Plant materials for use on surface-mined lands in arid and semi-arid regions. U.S. Department of Agriculture, Soil Conservation Service, SCS-TP-157, EPA-600/7-79-134.

U.S. Department of Agriculture, Soil Conservation Service. 1980. New Mexico State University and Colorado State University. Request for release, Hachita blue grama, a vigorous and drouth resistant blue grama.

U.S. Department of Agriculture, Soil Conservation Service. 1981. Land resource regions and major land resource areas of the United States. U.S. Department of Agriculture handbook 296.

U.S. Department of Agriculture, Soil Conservation Service, Los Lunas PMC. 1982. San Juan Coal Company field evaluation planting, Waterflow, New Mexico. Progress report.

U.S. Department of Agriculture, Soil Conservation Service, Los Lunas PMC. 1982. Sunbelt Coal Company, De-Na-Zin Mine field evaluation planting, Farmington, New Mexico. Progress report.

U.S. Department of Agriculture, Soil Conservation Service. 1983. New Mexico State University and Colorado State University. Agricultural Experiment Station. Release notice. Niner sideoats grama.

U.S. Department of Agriculture, Soil Conservation Service. 1984. National plant materials handbook, Sections 406 and 605.

U.S. Department of Commerce. 1941. Climate of the states - New Mexico. Agriculture yearbook separate no. 1849.

U.S. Department of Commerce. 1964. Climatic summary of the United States - Supplement for 1931 through 1952, New Mexico.

7

The Study of Soil Water
in Mine Reclamation

David G. Scholl

Introduction

This chapter discusses the pervasiveness of soil water as a factor in the reclamation of arid and semiarid lands. Soil-water relationships are at the root of many problems in revegetation and stabilization; thus, researchers should recognize and account for the status and behavior of water in spoils which are being reclaimed. The extent to which soil-water relationships are important to reclamation can be expressed with a few simple statements:

1. Successful revegetation depends foremost upon the growth of plants, and growing plants demand large quantities of soil water to satisfy their transpiration requirements.
2. Evaporation in arid and semiarid climates impairs revegetation by strongly competing with plants for the limited supply of soil water.
3. Soil water is the solvent for essential plant nutrients, which are carried to growing tissue in the transpiration stream. Vegetation wilts and dies in dry soil regardless of fertility.
4. Sodium, selenium, copper, and other potentially toxic elements can impede revegetation only if their *concentrations* exceed thresholds for plant tolerance. Sufficient quantities of soil water can reduce the harmful effect of these elements by diluting them or leaching them beyond the depth of plant roots.

There are many other ways in which soil water more subtly influences reclamation. Water content influences the aeration and temperature of soil, as well as its resistance to erosion by wind and water. Excess

135

soil water can leach undesirable soluble salts from fractured overburden materials into underlying aquifers.

Principles and Methods of Soil-water Research

The study of soil water must begin with an understanding of the cycle of water's movement into, through, and out of the soil. Figure 7.1 illustrates a hydrologic cycle for a reclaimed surface. Water arrives by precipitation and/or irrigation, but the amount of moisture which enters the soil may be less than the amount delivered. Water may be lost as evaporation before reaching the soil surface, or it may be lost as runoff, which happens when the delivery rate exceeds the soil's infiltration capacity. Runoff is a particularly undesirable avenue of water loss, because the water can pick up and carry sediments, causing sheet, rill, and gully erosion.

Irrigation is expensive, but it has the advantage that the timing, rate, and distribution can be tailored to the soils and to the needs of the plants used for revegetation. Irrigation will not cause runoff if it accommodates the capacity of the soil to take up water. On the other hand, water from precipitation can cause runoff and it can arrive before or after the growing season, when it is less valuable to plants.

Water which enters the soil can be lost by evaporation or by percolation to the water table, although the latter is infrequent in arid climates. The remaining water is potentially available for transpiration by plants, but that availability is qualified by numerous factors related to the condition of the soil. The study of factors influencing the availability of water to plants is called the study of *soil-plant water relations*. Following are various terms involved in the study of water's behavior at different points in its passage through the soil-plant-atmosphere continuum.

Porosity, structure, and water movement

Soil pores are a network of spaces between soil particles that may be filled with air or water. They vary in diameter from less than a micron to several millimeters (Brady 1974). The size distribution of soil pores depends upon the size distribution of the solid particles (the textural components: sand, silt, and clay) and upon the soil *structure*, which is related to the amount of cohesion between the particles. These soil properties dictate the rate at which water infiltrates into the soil and the rate of water's movement within the soil.

The sodium content *(sodicity)* of soils also influences the rate at which water infiltrates and migrates, because sodium ions reduce cohesion

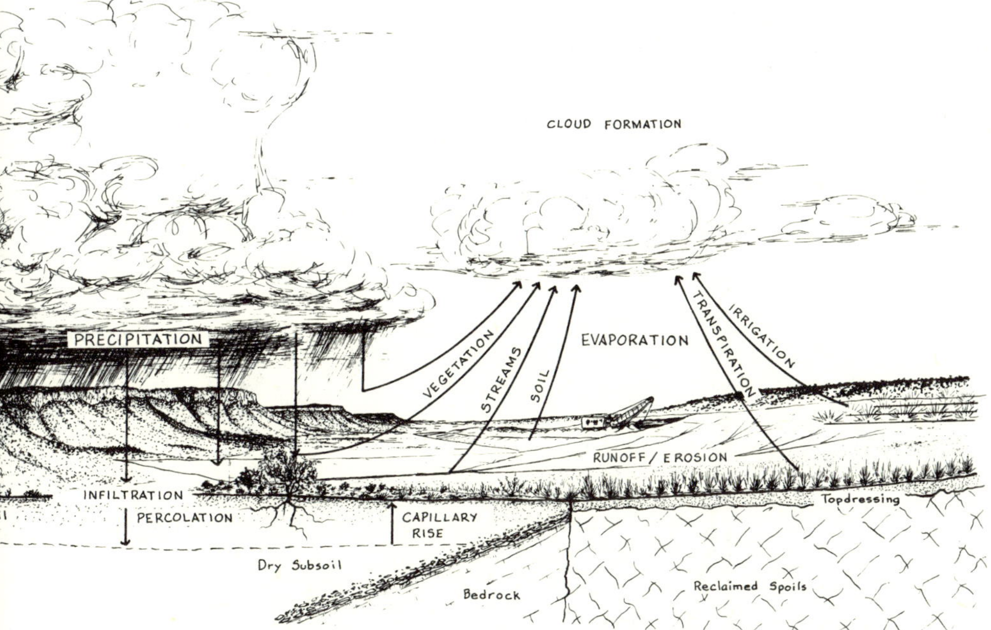

Figure 7.1.   A hydrologic cycle applicable to a reclaimed mine surface.

between soil particles and thereby reduce soil structure. Sodium ions, which have a single positive charge, line the negatively charged surfaces of clay particles, causing the individual soil particles to repel each other and become *dispersed.* This prevents the formation of larger pores which are important for the rapid infiltration or lateral movement of water in the soil. Calcium ions, which have a double positive charge, act oppositely, encouraging soil structure by causing particles to aggregate into larger units, which, in turn, produces larger pores.

These larger pores not only improve the *hydraulic conductivity* and water-storage capacity of soils, but they also allow for better root penetration. Still another advantage of large pores is to improve soil *aeration*, that is, the exchange of oxygen between the atmosphere and the soil pore space, where oxygen must be supplied to growing roots.

Water potential and availability

The loss of water from the leaves of a plant (transpiration) produces a *negative energy potential* within the plant. Water in the soil-plant-atmosphere continuum moves along an energy gradient (from high to low potentials), so water moves from soil pores into plant roots if the water potential of the soil is less negative than in the plant. The uptake of soil water by plants is resisted by the retention of water molecules

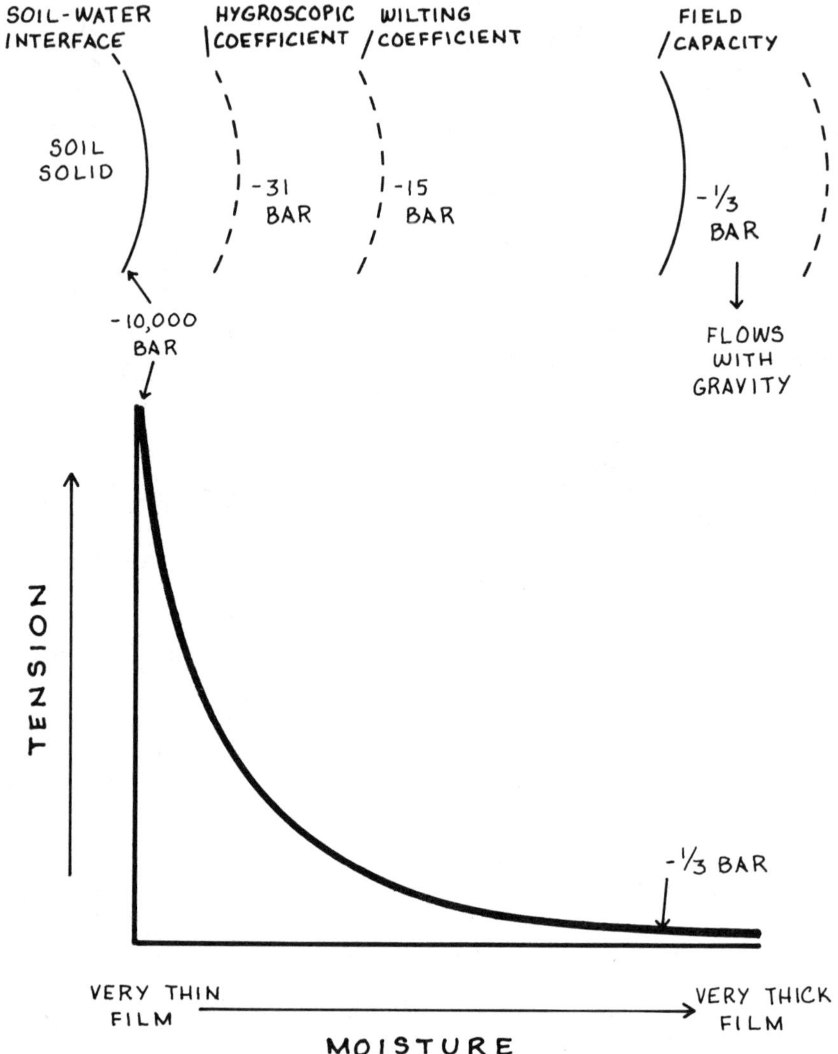

Figure 7.2. Capillary forces (adhesion and cohesion) in the soil pores (expressed as a negative pressure potential) increase exponentially as water is removed. If water is added to the soil, the water films will grow thicker and the negative water potential will drop. As the water potential goes below about one bar, the force of gravity will begin to dominate over capillarity, and water in the thick films will flow downward in the soil as drainage.

on the surfaces of soil particles *(adhesion)* and by the attraction between adjacent water molecules *(cohesion)*. As the film of water molecules lining the pores and adhering to particles becomes thinner, the amount of energy required to extract that water grows exponentially (Fig. 7.2). The result is that some of the water in the soil is unavailable to vegetation simply because the plants cannot exert sufficient pull.

Soils high in clay retain large amounts of water unavailable to plants. Figure 7.3 illustrates how clay retains more water at a given water potential than does sand, simply because the small size of clay particles results in an enormous surface area for the adhesion of water particles. The concentration of soluble salts in the soil also influences availability, since salts retain moisture osmotically. Saline soils have concentrations of salts which are sufficiently high to "compete" osmotically with vegetation, thus causing plants to wilt even when considerable water is present.

Field capacity and wilting point.   The strength with which water is retained in the soil (the *soil-water potential)* is described in bars or other units of negative pressure. Water available to plants in a loam soil (a soil with a relatively even mix of particle sizes) is generally held between ⅓ and 15 bars. If a soil is well drained, that is, if it is not underlain by a water-impermeable layer of clay or rock, then water held at less than ⅓ bars will drain off within about forty-eight hours after the soil has been saturated. After this excess water has drained, the soil is said to be at *field capacity*. Water can be taken up by plants until the soil-water potential is at 15 bars, which is considered the *permanent wilting point* (Klute 1965). The 15-bar value is typical for agronomic crop species, but species adapted to arid environments can tolerate somewhat more negative water potentials without wilting. In summary, water which is held in the soil between about ⅓ and 15 bars is available to plants, and is aptly called *available growth water*.

Methods and instrumentation in water relations

Precipitation, infiltration, and runoff.   In light of the numerous complex factors which influence the availability of water to plants, a rain gauge is not enough for studying water relations in reclamation. One must first know what fraction of precipitation actually enters the soil. A good estimate can be achieved by looking at the converse: the amount of water which runs off. This is often done on watersheds, which are instrumented with flumes that intercept and measure all departing water (see Chapter 2). However, flume data must be interpreted with a careful understanding of the watershed under study, because subsurface flow may resurface and be mistakenly counted as runoff.

Another approach is to measure the infiltration capacity of the soil. This may be done by placing an open-ended cylinder on the surface, pouring in water, and measuring the volume per unit time of water's disappearance into the soil. A better simulation, which accounts for rainfall effects, is to sprinkle water on a small bordered plot and collect all of the runoff for measurement. The sprinkler systems are sophisticated (and expensive) devices which can be adjusted to simulate various rainfall rates. All studies of infiltration rates must consider the moisture content of the soil prior to the experimental application of water.

Water content of soils. The traditional measure of the amount of water in the soil has been *gravimetric* analysis. In this procedure, a soil sample is weighed, dried, and reweighed, enabling calculation of the amount of water lost in drying. Gravimetric analyses are usually expressed as a percentage, computed by dividing the weight of water (the difference between weighings) by the dry weight of the soil. The principal liability of gravimetric analysis is that the soil samples must be extracted for weighing and drying, which can be difficult for deep layers and which disturbs the soil. A more recent method which gets around these problems uses a *neutron probe.* A small tube is excavated in the soil, into which the probe is inserted each time a reading is desired. The probe emits fast neutrons which are variously deflected by water molecules. The probe counts returning slow neutrons; these indicate the water content, since the more water molecules are in the soil, the more neutrons will be deflected back at the probe.

Water availability in soils. Gravimetric analysis and the neutron probe indicate how much water is in the soil, but they do not necessarily reveal the availability of that water. Availability can only be estimated based upon accompanying data regarding texture, porosity, and salt content. A *soil psychrometer* (Wiebe et al. 1971) measures the relative humidity of the soil atmosphere, which directly reflects the water potential, that is, the availability of soil water to plants. Small probes are buried in the soil with wires running to the surface. These wires are connected to a portable microvoltmeter whenever a reading of soil water potential is desired.

Perhaps the most direct measure of water availability is to study the plants themselves, whose water status is dictated by the soil. Psychrometers and other instruments are used to measure the water potentials in the leaves and shoots of vegetation.

Sampling strategies in studies of water relations

The important point in designing a study of water relations is to understand that not all of the water in the soil is available to plants. Further-

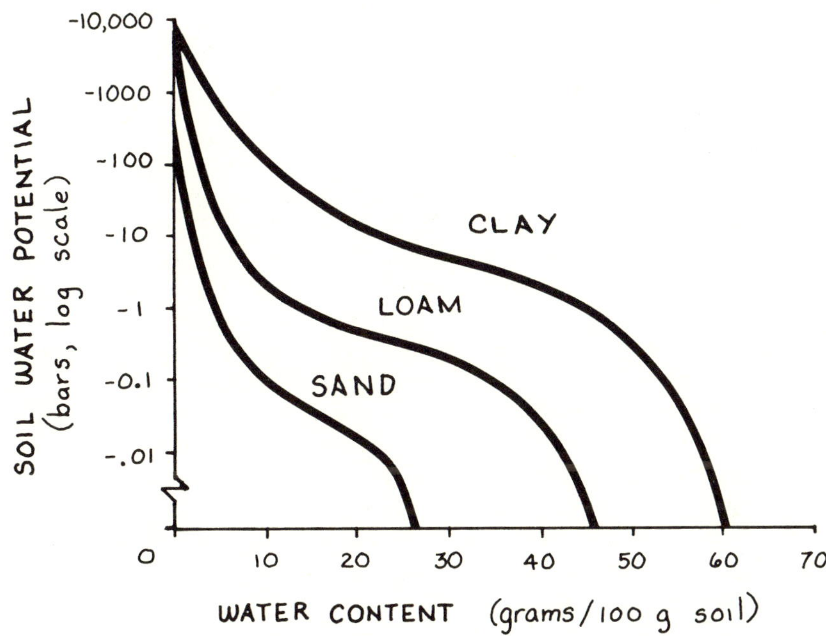

Figure 7.3.   Soil water extraction curves for three representative mineral soils. The lines show the relationship obtained by slowly drying completely saturated soils. (Modified from Brady 1974.)

more, not all the water which instrumentation indicates to be available (that is, water between ⅓ and 15 bars) is actually taken up by vegetation. For soil water to be usable, it must be in the root zone of plants during the growing season. Studies of water relations cannot be based on single measurements of water status in plants or soil. Sampling must be timed and located so as to provide the researcher with a complete picture of temporal and spatial patterns in the availability of soil water to vegetation.

Water-related Reclamation Research in the San Juan Basin

Water-related impediments to reclamation seem especially prevalent in the San Juan Basin because of the aridity of the climate and the adverse character of many of the soils and overburden strata which are encountered during mining. The relative lack of precipitation is compounded by the erratic timing of rain, which comes infrequently, but often torrentially. High winds and hot afternoon temperatures cause much water to be lost to evaporation.

Many of the soils and spoils in the Basin are high in clay and/or sodium, resulting in poor permeability to water. Sodium and other salts in these materials may be sufficiently high to inhibit plants osmotically, or to limit vegetation to highly salt-tolerant species. Furthermore, spoils on a single mine can be extremely variable in physical and chemical composition (Scholl 1982), making it difficult to develop remedial measures which apply to all the surface materials encountered in reclamation.

Most water-related research in the Basin began with the identification of a problem to reclamation, usually some aspect of the post-mining environment which limits the availability of water to vegetation. The next step was to propose and experimentally apply some remedial action, and to observe the effect of the action on soil-water potential. The final measure of the effectiveness of the remedial action is to observe the vegetation itself, sometimes in large-scale revegetation plots which demonstrate the long-term performance and desirability of the experimental procedure for reclamation. (See Chapter 4 for a detailed discussion of "demonstration areas.")

Moisture conservation practices

Moisture conservation systems have three goals: (1) to increase the amount of precipitation which enters the soil; (2) to increase the soil's capacity to retain water; and (3) to reduce loss of soil water to evaporation. Goals 1 and 3 may be achieved by mulching with an organic material such as straw or hay. Mulches are generally placed on the surface of the soil, where they reduce the impact of raindrops and provide shade (Fig. 7.4). This protection deters the formation of surface-sealing crusts and it prevents excessively hot soil temperatures, which cause high evaporation.

Organic mulches such as straw and woodchips can also be partially incorporated into the seedbed with a tractor-drawn disk. This protects the mulch from being blown away and increases the soil's capacity to retain moisture.

We studied the effects of incorporating straw and sawmill waste into sodic and nonsodic spoils at Mine C (Scholl and Pase 1984). Some of each material remained on the surface as mulch. Production of cool-season wheatgrasses was significantly higher on those treatments where mulch had been incorporated. We concluded that fine, well-decomposed organic materials are an excellent treatment for improving grass production when thoroughly mixed with the surface material at a rate of 5 to 10 tons per acre.

Figure 7.4. Barley straw (two tons per acre) broadcast on a mine-soil slope ready to be crimped or incorporated into the surface. The straw mulch will reduce evaporation, lower soil temperatures, and protect the soil surface from the erosive and compactive force of raindrops. The surface will remain porous and open for water to enter.

Gould et al. (1982) studied the effects of mulching in combination with irrigation at Mine A. Mulching produced no increase in the density of stands of native grasses, leading to a recommendation that mulch not be used where irrigation is adequate to obtain good germination. The authors did observe a favorable response by four-wing saltbush *(Atriplex canescens)* to the combination of irrigation and mulch. The long-term effects of mulch on the survival of plants after termination of irrigation was not studied.

Water harvesting

Water harvesting describes various means of concentrating water into areas where availability to growing plants is enhanced (Fig. 7.5). Chapter 3 describes how a high content of rocks in the soil can concentrate water into the remaining soil fines. It is also possible to take advantage of the natural impermeability of certain spoil materials, or to create

Figure 7.5.   Furrows which have trapped and held water just after a rainstorm. Newly seeded native grasses will germinate in the
furrows and produce an excellent stand.

impermeable surfaces with plastics and waxes, to intercept precipitation and channel runoff to areas for plant growth.

We tested the effectiveness of contour furrows for harvesting water
at Mine C (Scholl and Aldon 1978). A road grader was used to make
4-m- and 7-m-wide furrows which were left bare or which were treated
with a surface sealant such as paraffin or silicon. Runoff from the side
slopes was collected in 1-m-wide plant-growing areas at the center of
each furrow. Four native shrubs and a wheatgrass responded favorably
to the water-harvesting treatments compared to areas which had not
been furrowed. However, the surface sealants proved unnecessary at this
mine because the sodic spoils naturally shed water into the plant-growing
areas, indicating that spoil properties should be carefully evaluated before
investing in relatively expensive waxes or paraffins.

We also compared the performance of individual large contour furrows (0.2 m deep by 0.5 m wide, made with a single disk plow) versus
smaller, multiple furrows (0.1 m deep by 0.3 m wide) at Mine C (Scholl
and Pase 1984). The four multiple furrows were made by removing alter-
nate disk blades from the back gang of a heavy disk harrow, leaving the

remaining disks to create furrows. In this case, furrowing produced significantly better growth of wheatgrass *(Agropyron* spp.) when combined with the incorporation of partially decomposed pine bark into the surface material. Neither size of furrows increased production relative to the control treatments for the spoils which had not received pine bark. Measurements of soil-water contents among the treatments corresponded to our observations of wheatgrass production. We concluded that organic amendments may enhance the effectiveness of contour furrowing as a reclamation practice.

Terracing has been tested as a water-harvesting treatment at Mine A (Gould 1978, Ferraiuolo and Bokich 1982). Terraces were cut along the contour into a 10% slope of graded spoil, diked to retain water, and covered with a sandy topsoil material. The terraces were approximately 3.6 m wide and the harvesting slope above each terrace ranged between 3.3 to 6.6 m. The treatments received sprinkler irrigation for one year only, after which good stands of grasses and shrubs were maintained. Ferraiuolo and Bokich (1982) reported that plant covers and productivities in the growing areas were more than three times greater than in adjacent natural plant communities.

Depending on the properties of the spoils, water harvesting can be easily achieved by simply gouging the surface with pits or interrupting slopes with strategically placed dikes. Water-harvesting alternatives should always be considered in arid situations because irrigation is often expensive or impossible.

Irrigation

The essential purpose of irrigation is to establish seedlings (Fig. 7.6). Plants used for reclamation must survive without supplemental water after the first one or two growing seasons. If seeding rates are too high, much of the initial stand of vegetation dies after irrigation is withdrawn. More research is needed on the adjustment of seeding rates, water harvesting, and conservation measures to increase long-term plant survival on revegetated surfaces after irrigation is discontinued.

Studies so far have examined the amount and frequency of watering needed to establish seedlings. Most work in the San Juan Basin has been at Mine A, where initial studies indicated that native plants would be difficult to establish without irrigation (Gould 1978). Numerous combinations of rates and frequencies were tested there, with the most successful emergence achieved from the following combination: first, an initial "heavy" application of water; second, an application of 1.3 cm every other day; and third, after seedlings have emerged, applications

Figure 7.6.　Irrigation water from sprinklers establishes a wide diversity of native plants on formerly barren slopes. Straw, crimped into the surface, will help control erosion on the steep slope.

of 5 cm every two weeks for ten weeks. Stands receiving a total application of as little as 15 cm of water during the second growing season had higher survival rates than did stands with a single season of irrigation.

Wherever irrigation is used, studies such as the above must be conducted to determine the appropriate rate and frequency. Irrigation regimes require more site-specific adjustments to the local environment than do many other reclamation procedures. The choice of irrigation systems depends on the post-mining land use. Sprinklers are best if the goal is a fairly uniform cover of grasses and shrubs. However, a network of pipes with drip emitters is best if attempting to establish scattered shrubs on a steep slope with severe erosion hazard. Drip irrigation is widely used on steep (›60%) banks of tailings, where sprinklers would cause runoff and erosion.

Another important site-specific factor is water quality. Groundwater (or surface water in arid environments) can be saline or sodic, and can cause an accumulation of undesirable salts in soils on reclaimed surfaces. Weiler and Gould (1983) observed rising salinity in topsoil and reduced productivity at sites with saline irrigation at Mine A. Germination and growth of blue grama *(Bouteloua gracilis)* was completely inhibited by irrigation water which was both saline and sodic. On the other hand, four-wing saltbush grew reasonably well under the same irrigation regime, although emergence had been somewhat stunted.

Weiler and Gould (1983) also looked at the effect of saline irrigation on the distribution of salts in the soil. Salts in the irrigation water seriously deteriorated the upper 2.5 cm of surface material. A more insidious problem may occur when sodic spoil is mantled with only a few centimeters of topsoil or other nonsodic topdressing. High-quality irrigation water may dissolve sodium and other undesirable salts in the spoil, which rise via capillary movement to gradually contaminate the overlying material and reduce its favorability to vegetation.

Treatments to improve infiltration

Several properties of spoils and soils can cause low rates of water infiltration. These include high clay content, sodicity, and a repellency to water which is associated with residual coal particles in coal-mine spoils. Many shales in the San Juan Basin are plagued with one or more of these deterrents to infiltration (Scholl and Miyamoto 1984). Sulfuric acid, gypsum, and sand have all been tested for their ability to improve infiltration rates of water into spoils derived from shale (Miyamoto 1977, Scholl and Miyamoto 1983). Sulfuric acid, applied at 5 tons/ha, was the best treatment, with other measures showing little positive effect. Commercial wetting agents have also improved infiltration rates in spoils with high residual coal contents; however, the fine coal particles in these spoils cause considerable acidity (pH = 3.9) which deters plant growth. If wetting agents are combined with lime to counter acidity and if phosphorus is added to restore fertility, reasonable rates of infiltration and growth of vegetation can be achieved.

Summary and Future Needs

The pervasive problem of soil-water relations has been studied by researchers in government agencies, universities, mining companies, and consulting firms. This chapter has emphasized published findings; however, in-house corporate studies have been very important in obtaining the site-specific data required for effective water management at a given mine site. The ultimate goal for reclamation science, in general, and for individual mines, in particular, is to understand the behavior of water throughout the hydrologic cycle in reclaimed ecosystems. This goal is gradually being achieved as we understand the effect of reclamation practices such as water harvesting and irrigation at specific points in the water cycle.

Areas needing future research include: (1) applications of water-harvesting and conservation practices to prevent the regression of stands after irrigation is withdrawn; (2) improvements in water-harvesting and conservation practices for low-rainfall areas where irrigation water is not available; and 3) studies of mine-soil hydrology at the ecosystem level.

References

Brady, N. C. 1974. The nature and properties of soils. 8th edition. New York, NY: MacMillan Publishing Co.

Ferraiuolo, J. A., and J. C. Bokich. 1982. Irrigation, when is it necessary? *In* E. F. Aldon and W. R. Oaks, eds., Reclamation of mined lands in the Southwest, 170-173. Albuquerque, NM: New Mexico chapter, Soil Conservation Society of America.

Gould, W. L. 1978. Approach to reclamation of surface-mined land in the Fruitland Formation in New Mexico. *In* M. K. Wali, ed., Ecology and coal resource development, Vol. 2, 850-858. New York, NY: Pergamon Press.

Gould, W. L., J. A. Ferraiuolo, D. Leman, and R. Gabehart. 1982. Use of grass-hay or straw mulch for revegetation of minesoils under irrigated conditions. New Mexico State University Agricultural Experiment Station bulletin 484.

Klute, A. 1965. Water capacity. *In* C. A. Black, ed., Methods of soil analysis, Part I, 273-298. Madison, WI: American Society of Agronomy, Inc.

Miyamoto, S. 1977. Effects of wetting agents on water infiltration into water-repellent coal mine spoils. Soil Science 125:184-187.

Scholl, D. G. 1982. Properties of coal spoils of the Navajo Mine in northwestern New Mexico. *In* E. F. Aldon and W. R. Oaks, eds., Reclamation of mined lands in the Southwest, 35-42. Albuquerque, NM: New Mexico chapter, Soil Conservation Society of America.

Scholl, D. G., and E. F. Aldon. 1978. Water harvesting to establish perennial plants on semi-arid coal mine spoil. *In* M. K. Wali, ed., Ecology and coal resource development, Vol. 2, 735-739. New York, NY: Pergamon Press.

Scholl, D. G., and S. Miyamoto. 1983. Response of alkali sacaton and fourwing saltbush to various amendments on coal mine spoils from northwestern New Mexico. I. Acid spoil. Reclamation and Revegetation Research 2:227-236.

Scholl, D. G., and S. Miyamoto. 1984. Response of alkali sacaton and fourwing saltbush to various amendments on coal mine spoil from northwestern New Mexico. II. Sodic spoil. Reclamation and Revegetation Research 2:243-252.

Scholl, D. G., and C. P. Pase. 1984. Wheatgrass response to organic amendments and contour furrowing on coal mine spoil. Journal of Environmental Quality 13:479-482.

Wiebe, H. H., G. S. Campbell, W. H. Gardner, S. L. Rawlins, J. W. Cary, and R. W. Brown. 1971. Measurement of plant and soil water status. Utah Agricultural Experiment Station bulletin 484.
Weiler, G., and W. L. Gould. 1983. Establishment of blue grama and fourwing saltbush on coal mine spoil using saline groundwater. Journal of Range Management 36:712-717.

8

The Importance of Soil Ecology and the Ecosystem Perspective in Surface-Mine Reclamation

Walter G. Whitford and Ned Z. Elkins

Introduction

Attempts to reclaim surface-mined lands in the arid and semiarid western U.S. have been costly, ineffective, or both. Surface-mining regulations require that the areas be returned to levels of productivity equal to or better than that encountered on the area prior to mining operations. Mining companies desire to achieve this goal as quickly as possible in order to be released from the required bonding (to use those funds for new operations). Achieving high levels of primary (plant) production in a short period often requires that topsoil, fertilizer, and irrigation be applied, and that the surface be seeded with a mixture of native and exotic species. Even when such measures are employed, plant production may never reach the desired level, or production may not be sustained after the termination of irrigation and fertilization.

Applications of soil ecology to problems in reclamation

There are two basic problems in the reclamation of arid and semiarid lands which are addressed by soil ecology. First, soil ecology examines relationships between microfaunal and microfloral components within the soil system. Soil ecology also tries to establish the means and extent of interaction and interdependence between primary production and associated soil communities. However, only recently have the technologies and methodologies been developed which enable the holistic, ecological study of soil systems.

Second, soil ecology reveals ways to accelerate and improve the short-term success of revegetation by enhancing biological activity in the soil.

151

This idea is quite new (Parkinson 1978) because it has depended upon our first gaining an understanding of basic soil dynamics and processes from natural undisturbed ecosystems. In this way, soil ecology address-es the goal-oriented nature of mine reclamation, which has historically dealt almost exclusively with soil stabilization and quantitative prima-ry (plant) production. A company's financial responsibility for reclama-tion is terminated when it can demonstrate that primary production and vegetative cover on the mine site are statistically similar to a spe-cific pre-mining community (see Chapter 9). Thus, previous reclama-tion research has deemphasized soil ecology in favor of methods for growing and establishing plants using supplemental irrigation, fertili-zation, and other farming practices which supply and conserve nutri-ents and water.

Spoil characteristics in arid and semiarid lands

Spoil materials in arid and semiarid lands such as the San Juan Basin tend to have widely variable physical and chemical characteristics. Some materials are shales that weather into (water impermeable) fine clays. Spoils vary greatly in sodium content; sodic spoils require different rec-lamation strategies than nonsodic materials. Spoils from coal mining may have high levels of organic matter which is essentially residual coal. However, this organic material is not chemically or physically equivalent to soil organic matter. The texture of the spoil material is dependent upon the parent material and the degree of fractionation that occurs during backfilling and contouring operations. Even when com-pared to a badly overgrazed, deteriorated soil in adjacent areas, spoils are essentially lifeless.

Soil is a living system. To the uninitiated, soil appears to be a nonliv-ing matrix which serves to anchor plants and to provide water and nutri-ents to roots. These functions are physical and require no organisms. However, the nutrient supply function is dependent upon a complex aggregation of soil organisms. In the absence of such an aggregation of organisms, there is no organic matter decomposition, nitrogen fixa-tion, nitrification, or mineralization. Neither is there recycling of phos-phorus, nitrogen, potassium, and other nutrients in the zone of soil immediately adjacent to plant roots (called the *rhizosphere*). Native undisturbed soils also support a variety of organisms that modify the physical structure of a soil. For example, in mesic forest soils, earth-worms play a major role as effectors of soil quality because they turn over soil and organic matter and modify the structure of mineral aggre-gates. Although earthworms are absent in arid and semiarid soils, sub-terranean termites perform many of the same functions. Elkins (1983) found that subterranean termites enhance infiltration probably as a result of the numerous galleries (small tunnels) they make in the soil. They also move substantial quantities of soil from deep in the soil column to the surface and are a major factor in nitrogen cycling in arid ecosystems (Elkins 1983, Schaefer and Whitford 1981, Whitford et al. 1982b).

Native soils in the arid and semiarid regions of the southwestern U.S. are characterized by low levels of organic matter, which affect the status of both nutrients and water in the soil. The low organic content of these soils may, however, be a function of high rates of breakdown (decomposition) associated with soils in semiarid environments. Recent studies by Santos and Whitford (1981), Elkins and Whitford (1982), Whitford et al. (1981b) and Santos et al. (1984) have demonstrated that organic matter disappears at high rates in arid habitats during the growing season and that there is little correlation between rainfall and decomposition. The reported rates of decomposition are not necessarily paralleled by high rates of mineralization (Parker et al. 1984a); hence, arid and semiarid soils experience occasional "bottlenecks" in the availability of nutrients during wet periods. Soil organisms that are responsible for nutrient turnover processes are patchy in distribution and are generally associated with aggregations of litter around the bases of large plants (Santos et al. 1978, Freckman and Mankau 1977, West and Klemmedson 1978). Thus, plants in arid and semiarid landscapes can be described as islands of fertility surrounded by infertile interplant spaces.

If the landscapes to be reconstructed on surface-mined areas are to have a vegetative cover equal to or better than that present before mining, then one must first understand the entire soil biota and its network of interrelationships. Only then is it possible to develop a plan for the reestablishment of a comparable soil biota. Much of what we know of such biota and their interrelationships is based on studies in southern New Mexico. However, limited data on the soil biota in the San Juan Basin confirm the similarities of the biotas and processes (Elkins et al. 1984, Parker et al. 1984b, Ettershank et al. 1978a).

The Ecosystem Approach to Reclamation

The problems involved in reclamation of mined areas are best studied by using a whole ecosystem approach rather than the piecemeal approaches that have been taken to date. Using the ecosystem paradigm as the structural framework for designing research or selecting technologies for examination, we view plants, animals, microbiota, soil, and climate as interacting components in the transformation of energy and cycling of materials. The rates of energy flow and nutrient transforma-

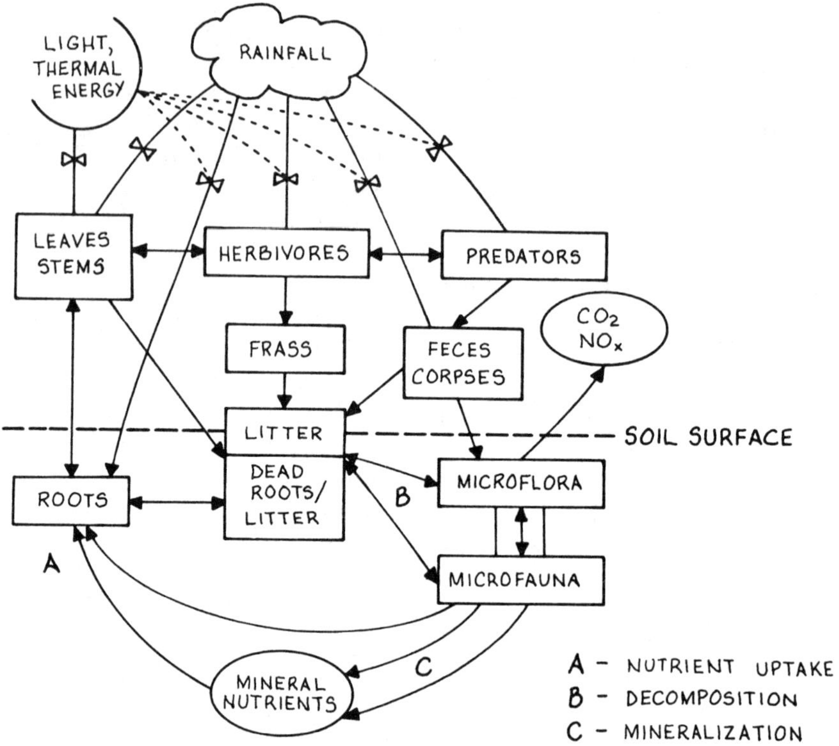

Figure 8.1. Structural and functional relationships of a generalized ecosystem. Arrows indicate flows of energy and nutrients plus feedbacks that control growth and population size.

tion are regulated by abiotic parameters (for instance, moisture, temperature, soil chemistry, soil structure) and by interactions among and between biotic elements (Fig. 8.1).

Using that basic paradigm, we can view the soil and vegetation as interacting subsystems of the entire ecosystem. Soil and vegetation receive inputs from some system components while they send outputs to other components (Fig. 8.2). Still other components impose rate-regulating activities on the soil-vegetation subsystem. In a stable ecosystem, the rates of energy flow and nutrient fluxes are maintained at rates that balance inputs and outputs. In order for equilibrium to be maintained according to this model, the internal rate regulators must

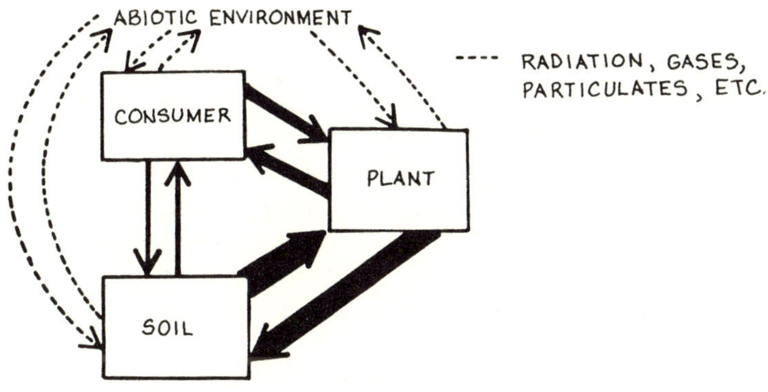

SOIL SUBSYSTEM

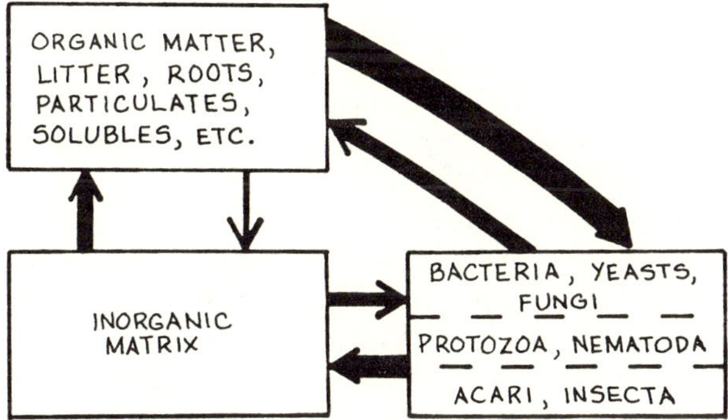

Figure 8.2. The relationships of the soil subsystem to other system components (upper panel) showing interactions. The lower panel shows relationships between biotic and abiotic elements within the soil subsystem. The width of the arrows is proportional to the exchange between compartments.

be in place. Hence, organisms in the soil subsystem that are important in decomposition and/or mineralization must be present and viable if the dynamic equilibrium of the system is to be achieved.

Ecological succession

The scientific basis for reclamation is ecological succession theory. The process of ecological succession is well documented in basic ecology texts. The reclamation problem, however, is not simply an ecological succession problem. It is one in which processes must be accelerated and the time scale must be compressed. Although raw spoils would undoubtedly undergo succession to some relatively stable assemblage of organisms within several centuries, we do not have the luxury of waiting that long.

A reasonable approach to the problem of succession on spoil materials is to apply our knowledge of the process to the development of techniques for accelerating succession. As described previously, soil is a complex *living* system, and actions that accelerate soil formation should accelerate the process of succession.

Ecosystem equilibrium

Surface mining destroys the natural ecosystem. The slow regeneration or succession of assemblages of organisms occurs in a system that is not in equilibrium. The ultimate goal of reclamation is to achieve an equilibrium state in the system. This need not involve the same species as were present prior to mining, *but* there must be species present that perform the same ecosystem processes. It is therefore essential to understand the ecosystem processes and types of organisms involved in these processes to design an effective reclamation strategy. It is also essential to apply our knowledge of ecological succession to determine how to more quickly achieve an equilibrium situation.

Stages of succession

Figure 8.3 depicts the processes involved in succession as well as the "stages" that we think occur on disturbed soil in semiarid areas. Pioneer plant species such as tumbleweed *(Salsola kali)* may germinate and establish even on raw spoils; however, the invasion is not uniform, resulting in patches of relatively dense vegetation interspersed among sparsely vegetated or open areas. Patches of plants disrupt air movement across the surface, producing eddy currents that result in the deposition of airborne particulates, including seeds of other plants, spores

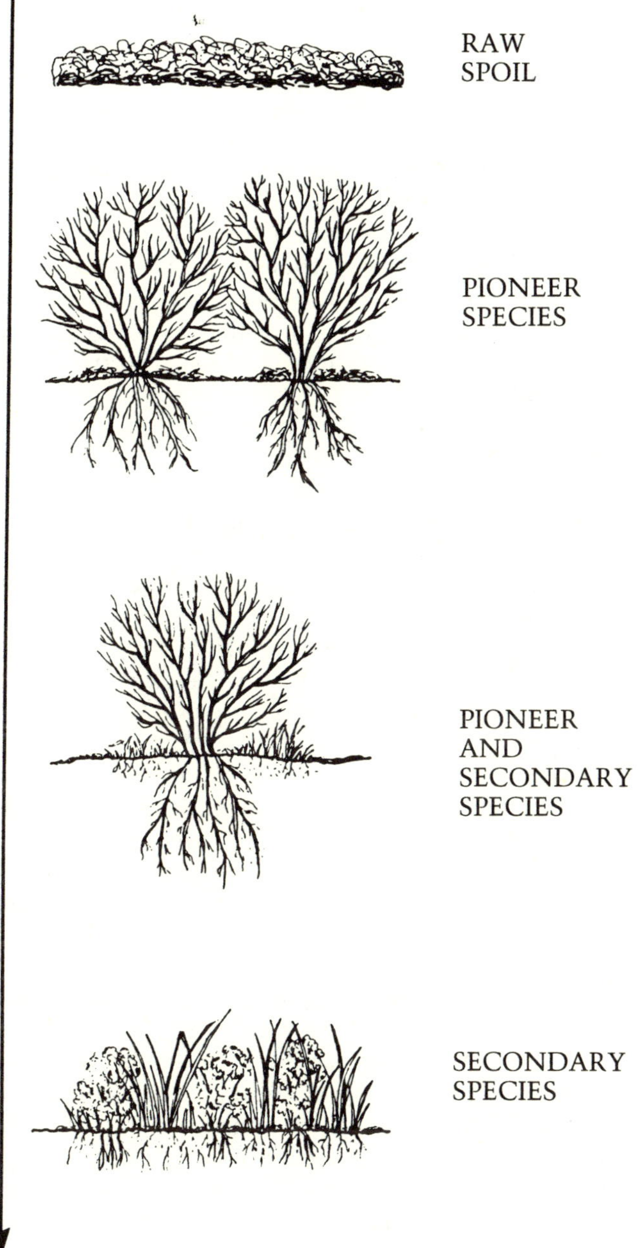

Figure 8.3. The stages that occur during ecological succession on mine spoils in the arid West. (The processess are described in detail in the text.)

of bacteria and fungi, cysts of protozoans, and dormant nematodes and microarthropods. These patches thus become islands where organic debris, seeds, soil particles, and the inocula of soil organisms become concentrated.

The pioneer species not only produce islands of deposition; they also modify the microclimate at the soil surface by shading, thereby reducing evaporation. Organic matter deposited around plants also cuts evaporation and contributes to changes in the chemistry and physical properties of superficial materials. The roots of the newly established pioneer species modify the inorganic soil material by providing a source of energy for rhizosphere microorganisms (bacteria, yeasts, and fungi) and by initiating channels that help to enhance aeration and the infiltration of water. Thus, the presence of a pioneer plant species initiates a series of events that gradually modifies the microclimate and the physical and chemical characteristics of the soil.

The rate of microclimatic modification depends upon the effectiveness of the pioneer species in producing these favorable "islands," the number and size of such islands, and the distribution of the islands across a disturbed landscape. As the soil environment is gradually modified, more sites have conditions which favor the germination and establishment of other species of plants. The pioneer species may temporarily serve as "nursemaid" plants for the next round of invading species. As each island becomes more diverse in structure and species composition, its effectiveness as an interceptor of soil particles and water increases, as does the rate of change of soil characteristics and ecosystem processes. Eventually, the interactions between plants, soil, microflora, and fauna will achieve some kind of dynamic equilibrium. That is, although the system will not remain static, the assemblages of plants and animals will remain essentially the same through time.

Accelerating the establishment of a soil biota

The above process describes how pioneer plants serve as foci for accumulation of inocula or gemmules of the soil biota. Spores of bacteria and fungi, protozoan cysts, nematodes, and mites are all transported by wind, as well as by insects and vertebrates. Thus, there is a potential for a relatively complete soil biota to develop soon after pioneer plant species establish. However, if the inorganic soil does not provide the necessary and suitable sources of energy and nutrients or is otherwise hostile, these organisms cannot establish. Even if topsoil, seed, and mulch are applied, the below-ground ecosystem requires longer than five years to approach the condition of the undisturbed system (Parker et al. 1984b). One method to modify spoils and enhance the development of soil biota is to mix in recalcitrant (decay resistant) organic matter (Parkinson et al. 1980), such as wood chips or bark. The addition of such organic matter provides an energy source for the soil biota that will be retained for several years. This allows the complex soil biota to survive and modify the material through their physiological and behavioral activities.

It is necessary to compare the composition and activity of the soil biota in reclaimed systems to that in undisturbed ecosystems in order to assess the desirability of a given organic amendment, an irrigation scheme, fertilization, topsoil application, or planting. Since soils are shallow and poorly developed in most areas of the San Juan Basin, it may be cheaper and more effective to modify spoils rather than to cover the material with salvaged "soil." Since the soil biota in native arid and semiarid soils is essentially limited to the surficial 10 to 20 cm, stockpiled soil will be devoid of a soil biota. Use of stockpiled soil or "borrow" soil, therefore, may not accelerate the development of a stable assemblage of plants and animals on a mined area.

The Soil Biota

Most reclamation research on soil-formation has considered only the microflora. Indeed, most soil scientists are only marginally aware of the fact that soil is a very complex subsystem, usually composed of many faunal components. Both field and laboratory studies of undisturbed soils have demonstrated the dependence of soil nutrient-cycling processes upon this fauna (Swift et al. 1979, Parker et al. 1984b, Santos and Whitford 1981, Cole et al. 1978, Anderson et al. 1979, Coleman et al. 1978). The soil subsystem processes are not unlike those of the aboveground system, but they are widely known. However, it is recognized that primary producers are very dependent upon a balance between decomposition and mineralization processes to provide the necessary nutrients for further vegetative growth. It is well documented that, in the absence of soil fauna, the microflora immobilize nutrients in their own biomass, thereby retaining nutrients in the soil and competing with plants (Swift et al. 1979, Parker et al. 1984b).

Since most of the San Juan Basin receives an annual average of less than 50 cm (20") of precipitation, one could logically ask, "Why are we concerned with nutrient processes when the factor limiting productivity is obviously water?" Such a statement is based on conventional wisdom, but not on the current evidence which suggests that nitrogen is far more important as a factor limiting long-term productivity in semiarid systems than is water (Floret et al. 1982, Penning de Vries and Djiteye 1982, Ettershank et al. 1978b, Whitford, unpublished data). In experiments where we supplemented the annual natural rainfall (230 mm) with 300 mm of water applied by an artificial rainfall system, plant biomass production was much greater on the watered plots than unwatered plots only when nitrogen was added. This indicated the importance of organisms involved in nitrogen cycling, and in other nutrient processes, to the success of reclamation programs in areas like the San Juan Basin.

Microorganisms

Bacteria and yeasts are known to be active in desert soils (Santos et al. 1981, Parker et al. 1984a), although their growth and activity are obviously limited by moisture. Because of their small size and rapid reproduction, they can grow in water films on the surfaces of roots and litter in dry soils. Experiments in which desert soils were kept wet by irrigation have shown that bacteria and yeasts are more limited by available carbon sources than by water (Parker et al. 1984a). In contrast to bacteria and yeasts, some soil microfungi are capable of growing even when overall soil moisture is very limited (Griffin 1981), that is, they are not confined to water films. Griffin (1972) found two genera of microfungi to be active at low water potentials; these genera are most frequently isolated from desert soils (Durrell and Shields 1960, Taylor 1979). Fungi are also more limited by substrate availability (carbon source) than by moisture. An unamended mine spoil would lack sufficient substrate to support these organisms regardless of the availability of water.

Mycorrhizae.   The majority of higher plants form a mutualistic relationship between certain groups of soil fungi and their roots. These fungus-root symbioses are referred to as *mycorrhizae*. Mycorrhizal associations have been shown to enhance the growth and survival of certain plants on mine spoils (Lambert and Cole 1980, Khan 1981). The establishment and maintenance of these relationships is recognized as being an important aspect of a reclamation program (Marx 1975, Reeves et al. 1979, Zak and Parkinson 1983). Of the types of mycorrhizae that exist, the ectomycorrhizal and vesicular-arbuscular (VA) types (Fig. 8.4) are the most common. Both kinds of mycorrhizae are important in nutrient uptake by plants, particularly phosphorus, and they may contribute to increased drought tolerance (Allen et al. 1981). For some plant species, mycorrhizal infection is obligatory, meaning that the species will not grow in the absence of the fungus.

Ectomycorrhizae are associated with woody shrubs and trees while the VA symbiosis is usually found associated with grasses and forbs. Ectomycorrhizae are characterized by a layer of fungal hyphae (called the *mantle)*, which covers the surface of feeder roots, and the occurrence of hyphae between and around the cortical cells (called the *Hartig*

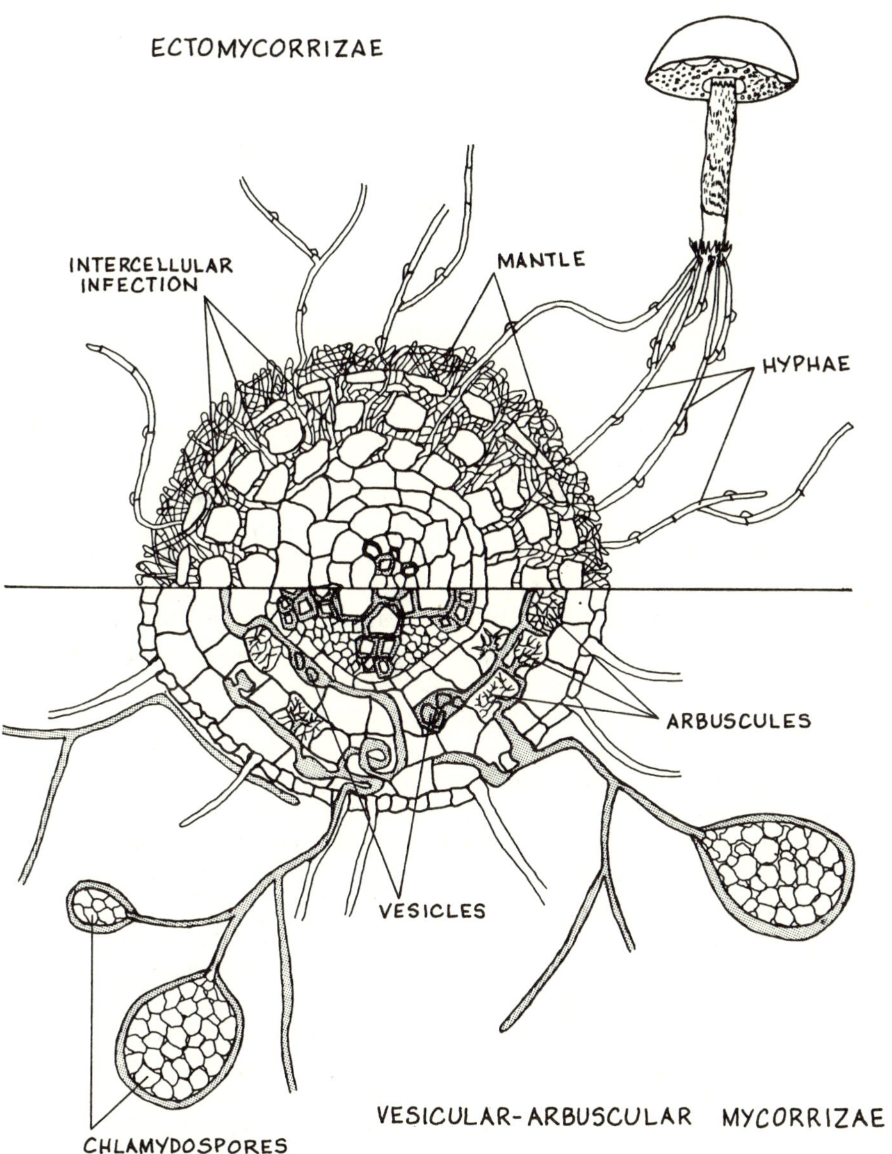

Figure 8.4. Diagram showing the relationships and structures of an ectomycorrhizal infection of a woody plant root (upper panel) and the relationships and structures of vesicular-arbuscular mycorrhizae and a grass root (lower panel).

net). In some instances, the symbiosis may change the morphology of the root. On the other hand, VA mycorrhizae reveal no external evidence of the symbiosis. The relationship is characterized by the occurrence of hyphae within and between cortical cells and the presence of vesicles and arbuscules within the cortical region of the root. The arbuscule, which is a finely branched structure that penetrates the cell membranes of the host, has been shown to be the site of nutrient transfer from fungus to host (Schoknecht and Hattingh 1976). Vesicles, which are large thick-walled bodies, can be either intra or intercellular and are believed to function as storage organs for the endophyte.

Most microorganisms are readily dispersed by the wind, sometimes in association with airborne soil particles. These include spores of most bacteria and fungi, although a particularly essential group of fungi, the VA mycorrhizae, lack an aerial means of dispersal. Many of the VA fungi produce soil-borne spores and/or sporocarps which are consumed by small mammals which eventually disperse the mycorrhizal spores in their feces (Trappe 1981). This mechanism of dispersal has also been reported for species of ectomycorrhizal fungi which produce their fruit-bodies underground (Maser et al. 1978).

Protozoans.   Protozoans are common in patches of organic matter in arid and semiarid soils. The predominate group of protozoans in these environments are naked amoebae (Parker et al. 1984b, and unpublished data), with ciliated forms only present in small numbers. Protozoans are an important link in the below-ground system because they can enter small pores and feed on bacteria in spaces inaccessible to the larger soil fauna (nematodes and microarthropods) (Elliot et al. 1980). The protozoans themselves are prey to nematodes in the larger soil spaces. Protozoa do require water films for their activity and will encyst in dry soils.

Nematodes.   Nematodes or roundworms are abundant and are especially important in the mineralization processes in soils of semiarid and arid ecosystems (Whitford et al. 1982a). The most abundant soil nematodes in these ecosystems are bacteriophagous; several genera of these have been identified by Freckman (unpublished) in soils of the Four Corners region. Freckman also found several fungivorous/omnivorous nematodes in these soils. Free-living nematodes make up the largest portion of the soil nematodes; the remainder are parasitic on plant roots. Most free-living nematodes in semiarid and arid ecosystems become *anhydrobiotic* when soils are dry (Freckman and Womersley 1983). Anhydrobiosis is a physiological state in which an organism dehydrates and reduces its metabolism to near-zero when water is absent. (In that state nematodes are inactive, desiccated, and with no measurable metabolism). When water becomes available again, the organism rehydrates, begins to feed, to reproduce, and to metabolize. This physiological capacity allows organisms to escape the dry conditions to which they are not otherwise adapted. Within a few hours after soils are wetted by a rain, the anhydrobiotic nematodes rehydrate and resume active feeding and reproduction.

Microarthropods.   Soil microarthropods are important in both decomposition and mineralization processes in semiarid ecosystems (Santos and Whitford 1981, Santos et al. 1984, Parker et al. 1984a). Soil microarthropods include the acari (mites), which are distant relatives of spiders, collembolans (springtails), and a variety of other, less abundant arthropods. The general morphological characteristics of these organisms are shown in Fig. 8.5. Among the acari three orders are important in soil processes: Prostigmata, Cryptostigmata, and Mesostigmata. Prostigmatids are usually most abundant in arid soils (Elkins et al. 1984), especially fungiphagous species.

Other relatively common prostigmatids graze on soil algae, yeast, and even bacteria. One group (Nanorchestidae) inhabits soils which are devoid of vascular plants (Santos et al. 1984). Other mites have been shown to be important in early stages of organic matter decomposition, possibly via regulation of nematode populations (Santos et al. 1981).

Cryptostigmatid (oribatid) mites are generally regarded as the most important regulators of decomposition and mineralization processes (Witkamp and Crossley 1966, Edwards et al. 1970, Crossley 1977). Two genera of these mites are relatively abundant in the Four Corners region. They may feed upon fungus, detritus, or both. The cryptostigmatids, however, generally feed only in surface litter; thus, they are not often found in association with decomposing roots or buried litter (Santos and Whitford 1981, Parker et al. 1984a).

Mesostigmatid mites prey upon soil acari and nematodes (Elkins and Whitford 1982). As has been discussed by Hanlon and Anderson (1979), Santos et al. (1981), and Swift et al. (1979), predators play an important role in regulating populations of nematodes and mites that graze on microorganisms. If the populations of grazers become too large, they can severely limit the growth of the primary decomposers.

Collembolans are wingless insects, some of which live in soil. They are the largest group of nonacarine soil microarthropods. We routinely find representatives of two to four families of collembolans in soil cores (Fig. 8.5). Several taxa of soil-dwelling collembola are known to enter a physiological state of anhydrobiosis or something similar in response to dry conditions (Greenslade 1981).

Termites.   In warm arid and semiarid regions of the world, termites are a numerically important component of the fauna, where they may serve many of the same functional roles in soil formation as do earthworms in more mesic environments (Lee and Wood 1971a, 1971b). The

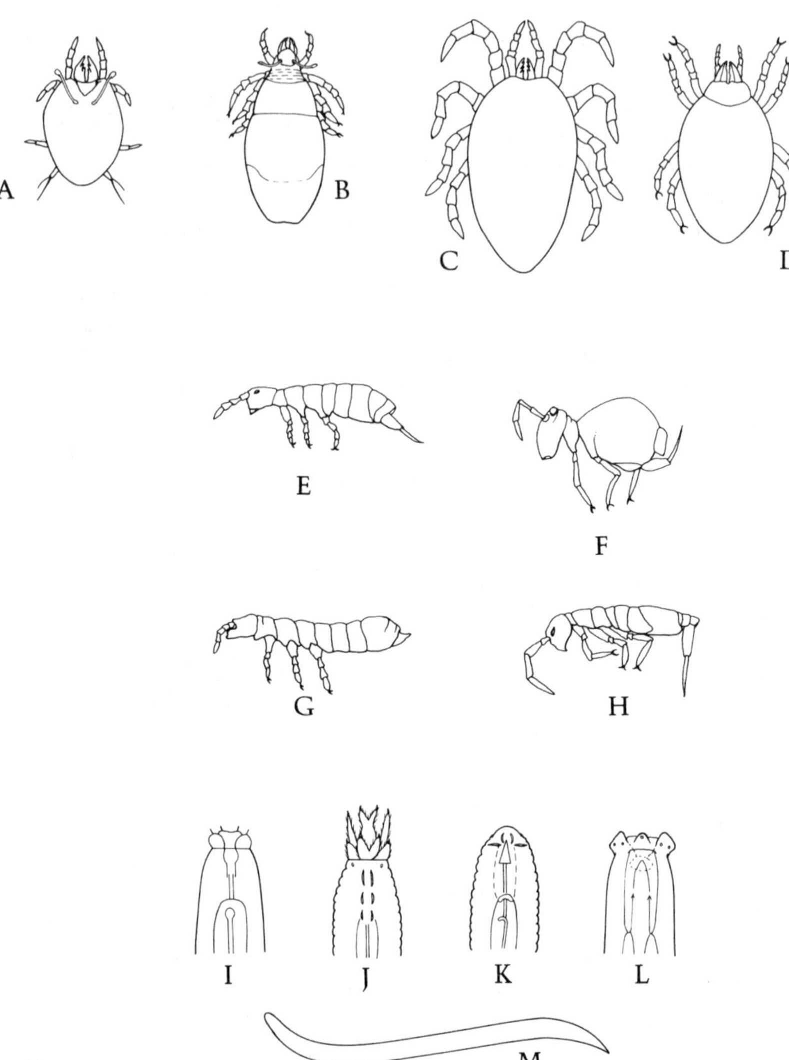

Figure 8.5. General morphological characteristics of several common soil acari, collembola, and nematodes that are found in the San Juan Basin. A. Fungus feeding tarsonemid mite, B. Fungus feeding-omnivorous oribatid mite, C. Predatory mesostigmatid mite, D. Omnivorous prostigmatid mite, E. Isotomid collembolan, F. Sminthurid collembolan, G. Hypogasturid collembolan, H. Entomobryid collembolan, I. Bacterial feeding nematode, J. Bacterial feeding nematode, K. plant parasite, fungal feeding, or predatory nematode, L. Omnivorous nematode, M. General body morphology of a nematode.

subterranean termite *Reticulitermes tibialis* is widespread in the Four Corners region. Ettershank et al. (1978a) reported population estimates of 185,000 termites per hectare (92,500/acre) in pinyon, juniper, and sagebrush habitats in the vicinity of Mine C. These numbers undoubtedly represent underestimates because they include only that fraction of a termite colony which is feeding upon a paper bait at the time of sampling. However, they convey the numerical importance of the termites.

Termites are social insects; a colony consists of one or more queens which lay eggs that develop into workers and soldiers. Soldiers make up a very small fraction of the colony in *Reticulitermes* spp. (Howard and Haverty 1980). Reproduction occurs via special male and female individuals that develop wings (alates) and that are nurtured by the workers. Mating occurs during synchronized flights by alates from all colonies in an area. Mating flights are triggered by environmental factors, generally a combination of rainfall and appropriate temperature.

We know little of the specific biology of *R. tibialis*, although some aspects of their biology can be inferred from our bait studies, laboratory studies, and work with other termites in arid regions. There are three characteristics of subterranean termites that make them particularly valuable in the development of soils from surface-mined materials: (1) they construct tunnels from their central colony chambers to food items on the surface; (2) they regularly ingest inorganic soil, apparently to provide nutrient salts for their symbiotic gut microflora; and (3) they use egesta (feces) mixed with soil particles to construct subterranean tunnels and surface structures over and around potential food materials. The tunnels markedly improve the capacity, rate, and depth of water infiltration into the soil (Elkins 1983). The tunneling and mixing of inorganic soil with organic matter (partially digested cellulosic compounds) also influences the mixing of soil profiles (Ettershank et al. 1978a). In a two-year study using laboratory setups (microcosms) to simulate conditions at Mine C, we found that *R. tibialis* survived longer than six months only where bark and wood chips were added to the top layer of spoil which had been covered with 5-cm of the mine's usual sandy topdressing. These microcosms were the only treatments where the entire soil profile had been mixed. Soil organic matter was increased throughout the profile, and total nitrogen was higher than in the other microcosms. Based on the work of Elkins (1983), Whitford et al. (1982b), Parker et al. (1982), and research at or around Mine C, it is clear that subterranean termites *Reticulitermes tibialis* are important members of the soil fauna in terms of their activities in undisturbed ecosystems and their potential to accelerate soil formation in mine spoils.

Interactions

Now that we have reviewed the major groups of organisms in the biota of arid and semiarid soils, we can examine their interrelationships and their contributions to processes in undisturbed ecosystems. This provides a background for the interpretation of existing data and the design of future research.

Decomposer organisms.   We shall consider the process of decomposition and the role of the soil biota in that process. The trophic (feeding) relationships of the soil biota are based on the literature and our experience with laboratory cultures. Surface litter (dead plant material) is processed differently from that buried in the soil. Since surface litter is subjected to rapid changes in temperature, the water content of the material changes on a daily basis. However, we have indirect evidence that certain fungi and microarthropods are active in surface litter in semiarid regions for at least a few hours each day, even after prolonged drought (Whitford et al. 1981a). On the other hand, neither protozoans nor nematodes are activated in dry litter during short daily episodes of "favorable" conditions (unpublished data and D. W. Freckman, personal communication).

Those species of fungi which can grow even in extremely dry conditions modify the litter to make it suitable for cryptostigmatid (oribatid) mites that feed on both the plant material and fungi. The most numerous microarthropods in dry litter are oribatids and some prostigmatids that seem to be feeding on the fungi. After the litter has been wet, the potential exists for other components of the soil biota—bacteria, yeasts, protozoans, and nematodes—to be active. This apparently varies seasonally (unpublished data), suggesting that seasonality may be as important as water and substrate for some of the soil biota.

Decomposition was slow relative to natural rates on mine spoils where microarthropods were absent (Elkins et al. 1984). Microarthropods affect decomposition by fragmenting plant material, ingesting materials that are later egested in the soil. They also transport fungal spores and bacteria on their exoskeletons and in their guts.

Unlike surface litter, roots and buried litter seem to have a characteristic sequence of invading organisms (Santos and Whitford 1981, Elkins and Whitford 1982) (see Fig. 8.6). The initial stages of decomposition are mediated by bacteria and yeasts. These are grazed by protozoans and nematodes which are, in turn, grazed by mites (Santos et al. 1981, Parker et al. 1984a). After 10 to 15% of the organic matter has disappeared, fungi colonize the decomposing roots and litter, and the soil biota shifts to a fungus-based food web. The primary grazers during this phase are fungiphagous nematodes and mites, the abundance of which is regulat-

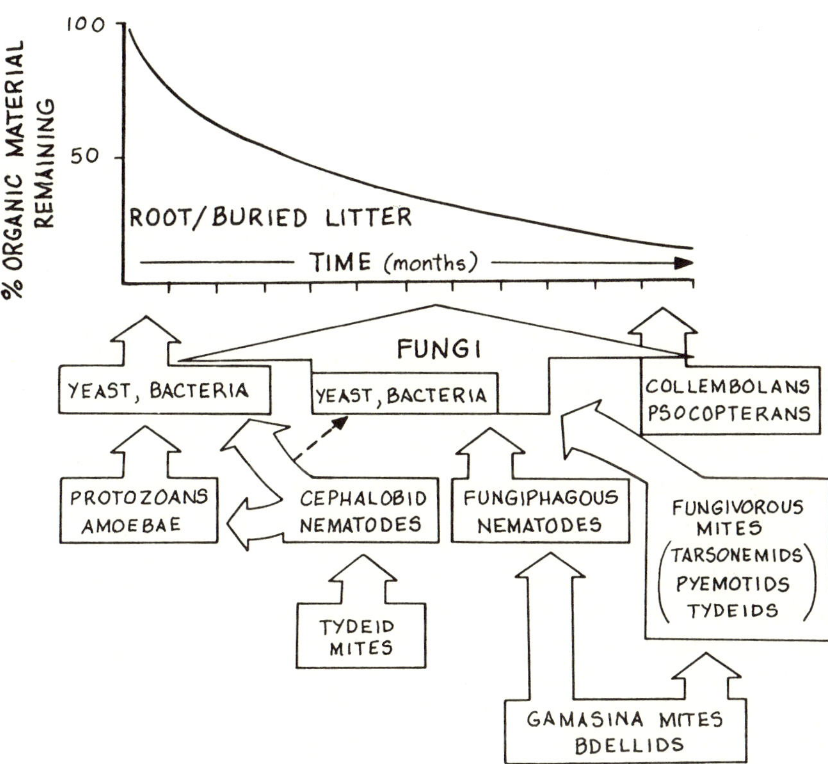

Figure 8.6. The relationship between stages of decomposition as shown by percentage of organic matter remaining and the soil biota that process buried litter and dead roots. The trophic relationships and relative quantities of transfers are indicated by the width of the arrows.

ed by certain predatory mites. In the late stages of decomposition, two insect groups—collembolans and psocopterans (booklice)—appear as part of the soil biota. They may eat the organic material itself or the fungi present on that material. Under some circumstances, these groups may exclude virtually all other groups of microarthropods. Maximum decomposition rates occur when there is a balance in the groups of soil biota; thus, rates are reduced if any groups are rare or absent.

Breakdown of organic matter is important in an ecosystem because it is the process by which mineral nutrients are released into the soil where they are again available for growth of new or established plants.

The relationship between breakdown of organic material and availability of nutrients is not necessarily a direct one-to-one relationship. Mineralization can be a very complex process for certain nutrients.

Nitrogen-cycling organisms.   From our studies, nitrogen (N) seems to be the principal limiting nutrient in arid and semiarid soils. For instance, consider the roots of a pioneer species, such as tumbleweed, which remain in a soil or mine spoil after the plant dies and is blown away. Fungi growing on these roots and utilizing the carbon within accumulate nitrogen from the surrounding soil. This process is referred to as *immobilization*. The fungi cannot obtain sufficient N from the dead root to synthesize the proteins and nucleotides necessary for the fungal growth. However, fungal hyphae in the soil can take up N from the soil solution. Thus, N from the surrounding soil is immobilized in the biomass of the fungi which are decomposing the dead roots. Unless that fungal biomass is grazed, the N remains there, unavailable to the rest of the system for new plant growth. We found (Parker et al. 1984a) that in the absence of microarthropods, N mineralization depended mostly upon the population densities of fungiphagous nematodes, whereas N mineralization depended upon both fungiphagous microarthropods and nematodes in intact systems. (See Fig. 8.7 for details of the role of soil fauna in nitrogen cycling in an arid system.)

In reclaimed areas, we found a relationship between N turnover and the degree of development of the soil biota (Parker et al. 1984b). It should be obvious from our description of soil biotic processes that, until a balanced soil biota is developed in a reclaimed area, there will be imbalances in nutrient cycling, reduced rates of decomposition, and reduced rates of nutrient turnover. A reduced rate of decomposition may be advantageous if it results in an accumulation of organic matter. However, the addition of a suitable organic material to the soil will encourage faster development of a soil biota and should result in the fastest reestablishment of nutrient cycling processes.

Methods Used in Studies of Soil Systems

As suggested in the previous section, we view reclamation research as an ecosystem reconstruction problem. In order to develop relatively stable assemblages of plants and animals on a mine spoil in arid or semiarid climates, it is widely recognized that plants must be used which are adapted to the local environment. It is less widely recognized that the soil subsystem must be functional in order to provide a reservoir of essential nutrients to plants and to provide the biological components necessary to make those nutrients available.

The potential effectiveness of revegetation procedures needs to be eval-

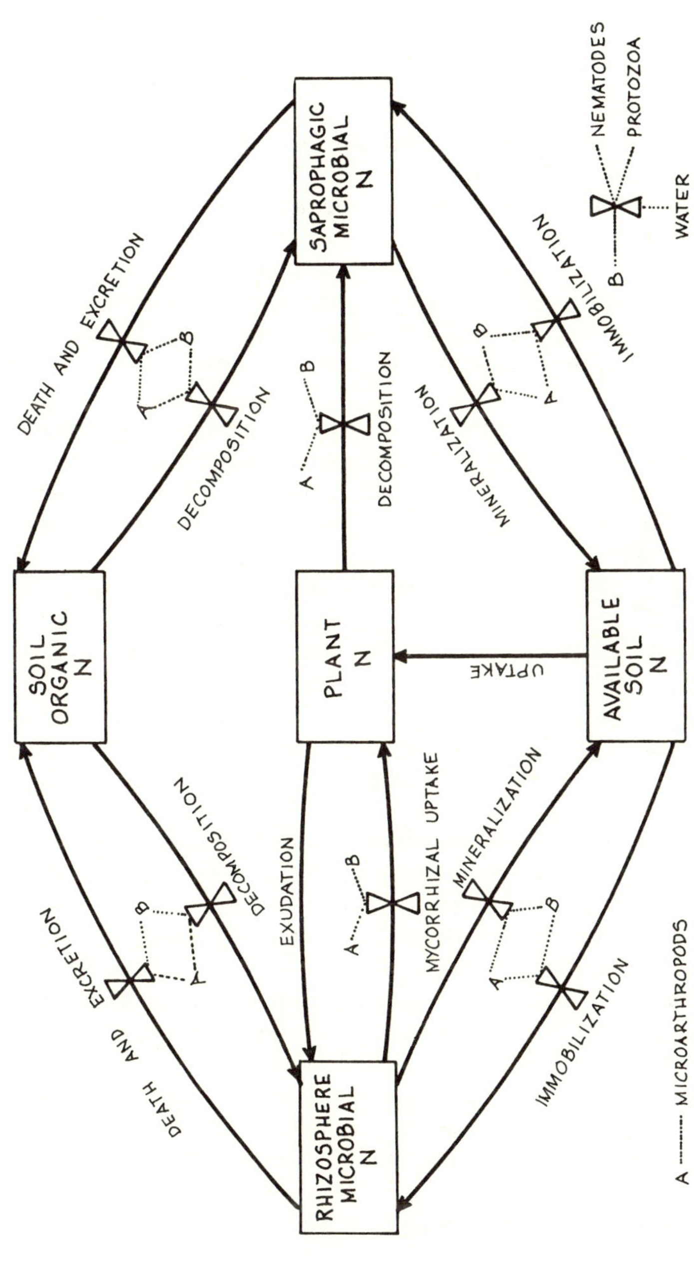

Figure 8.7. The relationships between nitrogen turnover in the soil subsystem and the soil biota, which regulate the nitrogen transformations. The symbols indicate areas of rate regulation. Wherever nematodes and protozoans are indicated, the soils must be wet for these organisms to be activated.

169

uated in terms of soil development and biotic activity. Our approach has focused on the soil subsystem in an attempt to evaluate the efficacy of organic amendments on spoils and to examine the rates of succession toward a "soil" when topdressing and barley straw mulch has been used (Elkins et al. 1984, Parker et al. 1984b). We have used both laboratory microcosms and field studies in our research.

Decomposition and nutrients

The basic processes in the cycling of nutrients are decomposition and mineralization, both of which depend upon numerous interactions among soil organisms. Fortunately, it is not necessary to account for all of the interactions in order to measure these processes.

There are many materials which can be measured before and after exposure to decomposer organisms (for example, filter paper strips, standard cotton strips, tagged lengths of straw, and litter bags), each with advantages and disadvantages. We prefer litter bags for the following reasons: (1) we can compare rates of decomposition between roots, aboveground plant material, or potential organic spoil amendments; (2) we can study litter bags on the surface or buried in spoil or soil, whereas most other techniques require burial; and (3) we can subsample litter bags to assess the development of the biological assemblages on various substrates. Litter bags are constructed simply and inexpensively from fiberglass window screen. We cut rectangles of fiberglass to achieve the desired size, fold the rectangle in half, and then run a large soldering iron down the sides to form an envelope closed on three sides. The bag is sealed with a soldering iron after the material is placed within.

Litter bags may also be used as baits to sample the soil biota. Since the soil biota is limited by energy sources as much as by the chemical or physical nature of the substratum, sterile litter bags may be buried in the soil, retrieved at desired intervals, and treated as sample units. The material may be extracted for nematodes or microarthropods and processed for bacteria, fungi, and protozoans, using techniques described in the following sections. We transport the litter bags to the field in individual plastic bags to collect and measure those fragments lost in handling. This is also done when bags are retrieved.

Bags returned from the field may be placed on Tullgren funnels for extraction (see the section below on soil microarthropods), or they may be processed directly. In either case, the material is oven-dried, weighed, and burned in a muffle furnace. This is necessary to correct for the infiltration of inorganic material into the bag. The ash weight and original weight are then used to calculate the amount of organic matter lost during field exposure; to do this, the following equation is used:

$$d = \frac{I + (A-Y)}{Si} - F; \; \%d = \frac{d \times 100}{I - Y}$$

(Elkins and Whitford 1982)

where:

d = estimated organic matter loss from each samples

I = initial dry weight of each sample (60° C for 72 h)

A = final inorganic weight (ash weight) of each collected sample (700° C for 4 h)

Y = weight of the initial inorganic content of each initial dry weight (estimated from each of ten standard samples of each treatment)

Si = estimated inorganic content of the soil (mean of ten samples of soil burned in a muffle furnace)

F = final dry weight of each sample (60° C for 72 h) after collection

Rates of organic matter loss may be compared across treatments using first order rate constants or other exponential decay models (Mellilo et al. 1982).

Microorganisms and protozoans

Numbers of microorganisms and protozoans are estimated from sub-samples taken from soil cores. To estimate the numbers of microorganisms, a 1.0 g sample of soil is first mixed in a blender with 100 ml of sterile water for three minutes. Ten ul of this mixture is then pipetted into two circular areas with 1-cm diameters on a microscope slide. The circular areas are delineated by placing clear adhesive tape (Scotch tape) on the microscope slide and removing a 1-cm diameter disk with a cork borer (Fig. 8.8). The soil smear is left to air-dry and then is heat-fixed by passing the slide over an open flame for approximately three seconds. The soil smears are stained with a fluorescent solution and washed to remove excess. The fields are prepared for counting by application of glycerol which is allowed to settle. For fungi we use an ocular micrometer grid and count the numbers of intersections. We count thirty fields per soil smear. For bacteria we count ten fields. The fungal hyphae and bacteria fluoresce with the dye when examined under a microscope with epifluorescence. Counts and diameter measurements are converted to biovolumes to obtain biomass estimates (Babuik and Paul 1970, Jenkinson and Powlsen 1976, Jones and Mollison 1948, Van Veen and Paul 1979).

Protozoans are estimated by the most probable number (MPN) method (Cutler 1970, Singh 1946). One gram of soil or litter is mixed with 100 ml of 0.5% NaCl solution in a blender for twenty seconds and dilut-

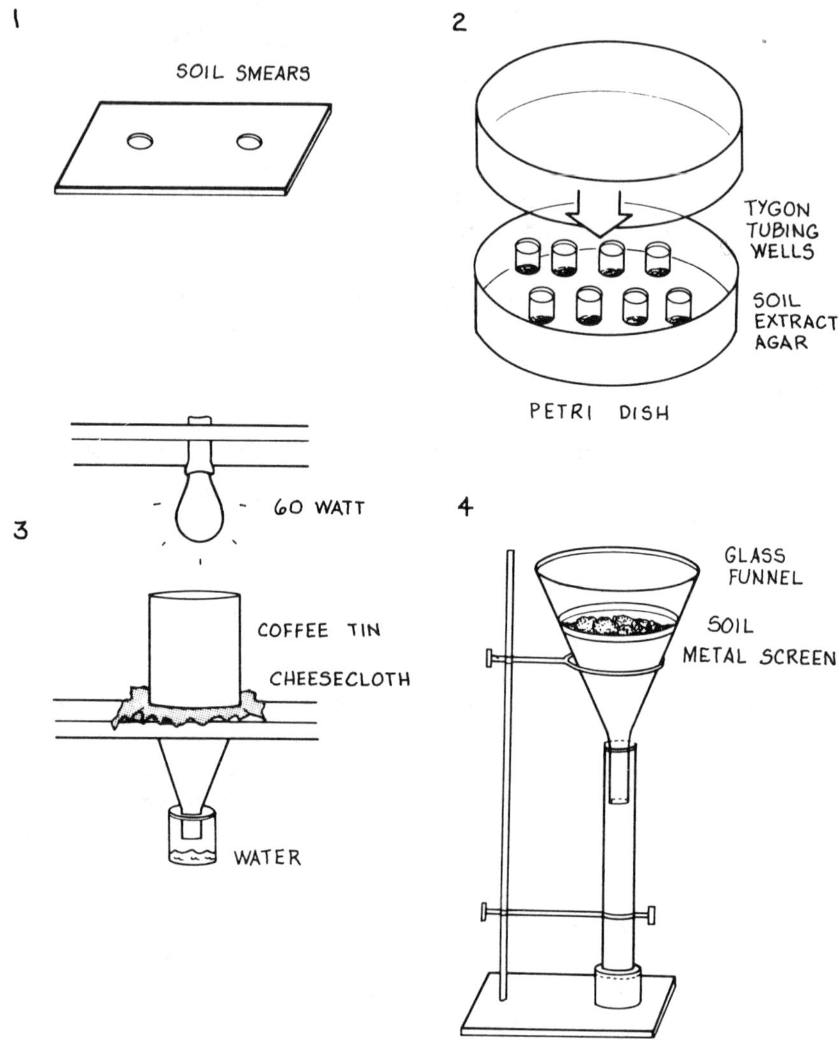

Figure 8.8. Examples of materials used in soil ecosystem studies: (1) microscope slide with "fields" for soil smears containing bacteria and fungi, (2) most probable number plate for protozoans, (3) Tullgren funnel extractor for soil and litter microarthropods, and (4) Baerman funnel extractor for soil and litter nematodes.

ed four times. The dilutions are shaken vigorously, and 0.1 ml of each is plated into five 1.5-cm diameter rings on the MPN plate. The MPN plates are made by pouring approximately 20 ml of soil extract agar into petri dishes. The MPN rings are cut from Tygon plastic tubing and imbeded in the agar before it sets (Fig. 8.8).

Microarthropods and termites

Santos et al. (1978) and Franco et al. (1979) have found that unvegetated areas in two desert habitats were characterized by only one or two taxa of microarthropods. These authors suggested that the distribution patterns of microarthropods reflect patterns in the distribution of litter and vegetation. Because of the widely known relationships between microarthropods and plants in arid habitats, we use a stratified sampling procedure for soil microarthropods. If we are sampling under shrubs, we take core samples under plants at ½ the canopy radius. If we are sampling under grasses and/or forbs, we take cores at the edge of the canopy or at a location which includes a fixed portion of the canopy. We use a 4.5-cm diameter plastic corer when sampling in sandy soils and a 4.5-cm diameter stainless steel corer when sampling in clay or heavy loam soils. Samples to 10-cm depth are adequate for most soil microarthropods. As in most soil biology, the number of samples taken is a compromise between the ideal and what can handled realistically. We have found that, with stratified sampling, eight to twelve cores provide an adequate sampling of the soil microarthropod populations.

We have experimented with a variety of techniques (Phillipson 1971) and have found that a modification of a Tullgren gradient extractor works best for samples from arid and semiarid regions. That extractor can be constructed simply and inexpensively by using readily available materials (Fig. 8.8). Insects are extracted by using water as opposed to alcohol or picric acid, because we have extracted little or nothing when using the latter two reagents. The water provides a humidity gradient as well as the thermal gradient. The microarthropods remain on the surface of the water, because of the surface tension, where they are more easily counted and identified before transfer into a preservative.

Subterranean termites can only be sampled indirectly. We utilize a technique developed by Haverty et al. (1974) in which subterranean termites are sampled in baits laid out in a grid on the sampling area. The baits are standard toilet paper rolls covered with aluminum foil except for the bottom of the roll which contacts the soil surface. The rolls are spaced at 1 to 2 m distance. At periodic intervals (two weeks minimum between samplings) the rolls are lifted, and termites, if present, are shaken onto a pan for counting. Several soldiers, if present, should be preserved with 70% ethanol because soldiers are needed in order to identify subterranean termites to species (Wiesner 1965).

Nematodes

Freckman (1982) recently reviewed nematode contributions to ecosystems and emphasized the importance of adequate sampling for obtaining reliable data. As with microarthropods, we suggest a stratified sampling scheme for nematodes. In arid ecosystems, nematode populations are higher under shrubs and grass clumps than in open spaces (Freckman and Mankau 1977). When sampling at surface mines, one should sample near or below grass clumps and shrub canopies to ensure comparability. Goodell and Ferris (1980) showed that it is more economical and more efficient to take several soil cores and combine them into a single sample rather than use one core as a sample. If using standard soil corers (diameter 2 cm), it is advisable to take four to five cores around the base of a plant or within a 100 to 200 cm^2 area. These can then be bulked to form a single sample. Sampling depth should be determined for each area based on the expected depth of penetration of roots and water into the soils. In most soil studies, the number of samples depends more upon the time available for extraction, counting, and identification, rather than upon a statistical determination. We have found that six to ten samples is minimal for encountering the usual variation in soil nematode communities in arid systems.

The frequency of sampling will be determined by the objectives and target questions of the study. If comparisons between treatments (for example, organic amendments) are to be made, two to three sampling periods within the growing season will probably suffice. However, if one is examining rates of population growth of predatory nematodes in response to a perturbation such as irrigation or fertilization, sampling frequency needs to be weekly or every two weeks because of the short development time of some species.

There are a number of extraction techniques, most of which use water to separate nematodes from soil or litter bags. We have found that a combination of the Cobb sieving technique and the Oostenbrink cotton filter provides fast, clean samples for analysis (Nicholas 1975). Methods which require a long-exposure time to water reveal potential populations, not actual populations, because eggs of some species hatch within forty-eight hours (Freckman 1982). If the sample is plant material, it must first be fragmented and mixed with water. We fragment plant samples in a Waring Blender prior to extraction.

In comparative studies, the Baermann funnel method requires the least effort and time. In this technique, the soils are placed on a layer of one-ply paper towels supported on a wire screen (Fig. 8.8) in a 100-cm-diameter glass funnel. The funnel is filled with tap water to the bottom of the screen and nematodes removed after forty-eight hours.

In the sugar flotation technique described by Freckman et al. (1975), the soil sample is stirred into a concentrated sugar solution containing an antiflocculant. The liquid containing the nematodes is then decanted through a 325-mesh sieve to collect the nematodes. These are then washed onto a 400-mesh sieve and transferred to a preservative before counting. To facilitate counting, nematodes are transferred to petri plates or watch glasses and counted under a dissecting scope at 10x magnification. Subsamples are placed on slides and examined under a compound microscope to allow identification to genus or species. Organisms can be assigned to trophic groups based simply on general head and esophageal structures (Fig. 8.5).

Developing expertise

From our description of the soil biota and their interrelationships, it should be obvious that a team of scientists is required to study at the ecosystem, or even the (soil) subsystem level of organization. Such an effort must be managed by an individual with an interest in the general problem and a willingness to develop a basic knowledge of all parts of the system. It is also important that all members of the research team be committed to solving the overall problem. In our work, we have assembled a team of workers who have the responsibility for various parts of the study. However, we still require the assistance of taxonomic specialists to identify the various groups of soil biota. However, even when taxonomic specialists are unavailable, that should not deter researchers from tackling soil ecosystem processes and problems. For example, in our studies so far, we have not found it necessary to classify fungi, bacteria, protozoans, nematodes, or microarthropods to the species level in order to understand the system processes which are important in reclamation.

How can one learn enough about soil organisms to begin studies? There are several general references that will enable one to begin to work on soil organisms. For soil acari, we have found that Krantz (1978) provides a step-by-step introduction to the organisms. Another valuable book (Page et al. 1982) covers techniques for most chemical and microbiological analyses of soils, and it has a comprehensive list of references. However, no book can substitute for spending a few days in a laboratory working with someone who has experience with the taxonomic group of interest.

Soil Ecosystem Studies of Reclamation in the San Juan Basin

The San Juan Basin in northwestern New Mexico affords a highly diverse, often unique environmental setting for the study and implementation of reclamation technologies. The Basin is not easily reclaimed by using standard practices because of the geologic complexity of potential overburden materials, the limited availability of topsoil, and the lack of precipitation (or groundwater for irrigation). Innovative technologies are needed for reclamation to succeed in the Basin and elsewhere in the arid Southwest.

Only recently has there been an increased awareness of the interactive roles of soil and litter organisms in the decomposition of organic matter and the cycling of nutrients. Ecosystem stability and long-term productivity depend upon the establishment of a stable decomposer subsystem. Research on the soil biota has been conducted with the goal of evaluating current reclamation practices (for example, fertilization, mulching, and irrigation) and suggesting new practices to mining companies and land management agencies.

Some soil-related research has already been conducted in the Basin, but there have been limitations in the early development of this effort, including a lack of sufficient funding. Local mining companies have funded many reclamation research projects, but mostly projects with immediate goals and practical value, for instance, comparisons of various schemes for irrigation, mulching, or cultivation.

Perhaps the best approach for the study of soil organisms and their role in mine reclamation has been to utilize existing reclamation test plots. This approach has several advantages: (1) there are numerous such test plot areas in the region, some active and some abandoned; (2) these plots exhibit a wide range of treatments, including various combinations of topsoil, fertilizer, irrigation, mulches, and seedbed preparation techniques; (3) the plots occur in a wide range of environmental settings, with different elevations, topography, native vegetation types, spoil characteristics, and climates; (4) the plots may be quite old, enabling researchers to address long-term processes in the soil; (5) the plots may be studied immediately and inexpensively, compared to projects where the plots must first be established; and (6) such studies are generally approved and supported by the mining companies and the researchers who initiated the plots.

Studies of organic amendments at Mine C

An example of this type of "piggyback" research was performed between 1978 and 1982 at Mine C. A multiple plot study, addressing

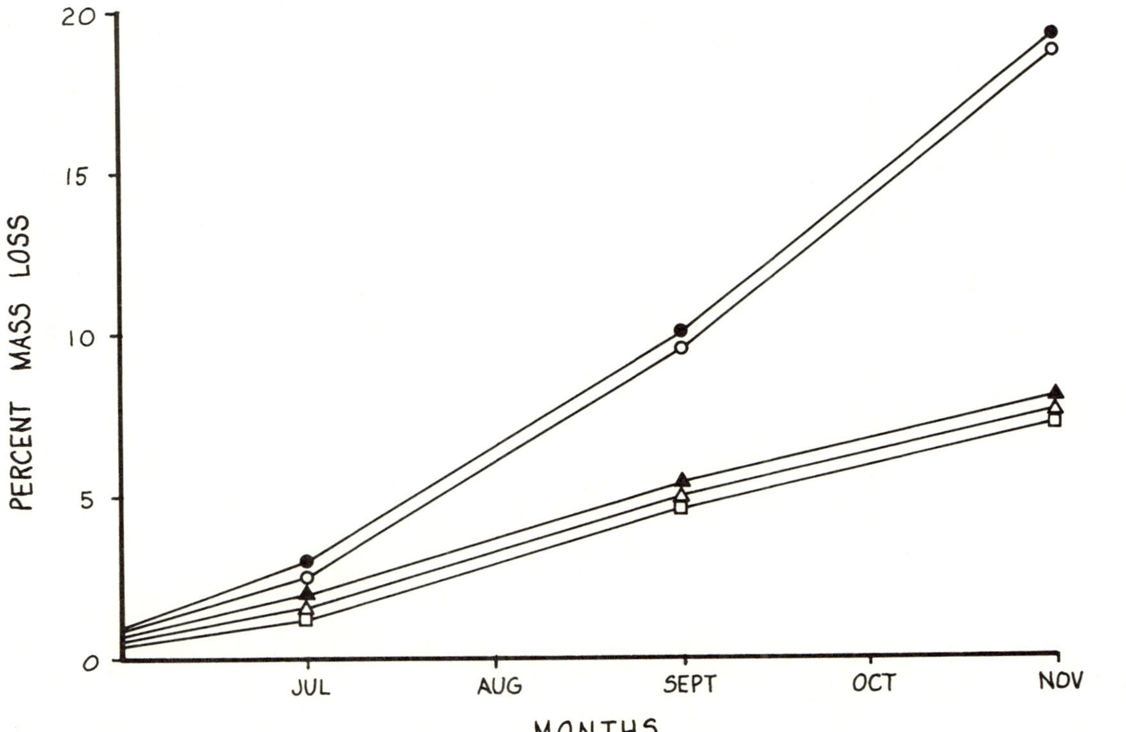

Figure 8.9. The relationship between organic amendments mixed with raw spoils and decomposition of barley straw at Mine C in northwestern New Mexico. The comparisons ofimportance are unmined soil and raw, unmined spoil.

the responses of certain wheatgrasses to various organic amendments and contour furrowing, was initiated on regraded mined lands during 1978 by the Rocky Mountain Forest and Range Experiment Station (Scholl and Pase, 1983). In 1980, we began a broadly designed ecological study of the associated soil/spoil systems on each treatment plot (Elkins et al. 1984). Our studies were designed to demonstrate that an ecologically sound reclamation procedure should exhibit an increased biotic potential, coupled with a high rate of litter-processing. Further, we sought to assess the relative ecological success of the various procedures in the original study by evaluating the responses of the total soil biota (microarthropods, nematodes, protozoans, fungi, and bacteria). The following treatments were compared: raw spoil, bark-amended spoil, straw-amended spoil, topsoiled spoil, and an adjacent unmined control area.

We used litterbags to compare the rates of decomposition of straw at the different treatments. Nematodes, protozoa, and microarthropods were extracted using the procedures described above. Throughout the study, bark-amended spoils demonstrated the highest rate of decomposition among the treatments, with rates that were similar to those at the adjacent control area (Fig. 8.9).

Tables 8.1 and 8.2 present statistical comparisons of the extracted or estimated numbers of soil organisms from the different treatments. Microarthropods were most abundant and diverse in the bark-amended spoils, again with values similar to those at the control site. Plots receiving straw or topsoil had many fewer microarthropods. Since these organisms live longer and respond less rapidly to temperature and moisture fluctuations than do soil microbes (Whitford et al. 1981a), the results suggest that bark-amended spoils are functioning similarly to native soils from a systems perspective. The reduced activity of decomposers in raw spoil, topsoiled spoil, and straw-amended spoil indicated that nutrient cycling was relatively slow and inefficient in those treatments.

Decomposition rates and microarthropod activity differed significantly between the two organic amendments, with bark exceeding straw in both cases. The difference may owe somewhat to significant differences in the rates at which the amendments were applied (18 tons/ha for bark, 5 tons/ha for barley straw), as well as to differences in the chemical and physical nature of the two substrates (for example, C/N ratios, lignified carbon content, and so forth). Neither the primary productivity (of plants) nor the activity of soil microbes differed between the two amendments.

The lack of difference in decomposition rates among the straw-amended spoil, topsoiled spoil, and unamended spoils suggests that neither straw nor topsoil are effective sources of energy or nutrients for the decomposer organisms (microflora and microfauna). The microfauna are especially important as regulators of the rate of decomposition and mineralization

Table 8.1. Average numbers of bacteria and protozoans and mean hyphal length of fungi in barley straw buried in litterbags at 10-cm depth in amended mine spoils and unmined soils.[*]

Treatment	July 28	Sept 3	Oct 24	
Topsoil	3.8a	13a,b	14a,b	
Bark	3.1a	18b	18b	BACTERIA
Straw	3.6a	5.5a	19b	(mean number x 10^9/bag)
Unmined	4.1a	5.4a	7.0a	
Topsoil	9.1a	250b	800a	
Bark	100b	840b	910a	PROTOZOANS
Straw	25b	57a,b	830a	(mean number x 10^6/bag)
Unmined	8.2a	4.5a	100a	
Topsoil	2.2a	18a,b	18a	
Bark	6.0a	28b	26a	FUNGI
Straw	3.8a	12a	23a	(hyphal length x 10^4[m/bag])
Unmined	3.8a	9.2a	18a	

[*]Numbers followed by the same letter are not significantly different at P .05.

(Santos and Whitford 1981). Three taxa of mites (tydeidae, torsonemidae, and pyemotidae) which are considered important in the regulation of decomposition and mineralization were well represented in the bark-amended spoil and native soil, but were generally rare in the other treatments.

Estimations of biomass or other measures of the abundance of microflora do not necessarily reflect the biological *activity* of those organisms. Our studies of decomposition rates indicated that microflora (fungi and bacteria) were less active on straw-amended spoil, topsoiled spoil, and raw spoil than on the other two treatments, even though the numbers of organisms in microbial populations did not reflect these differences.

Decomposition and mineralization are enhanced by the grazing of microflora by microfauna (Nicholas and Viswanathan 1975, Coleman et al. 1978, Anderson et al. 1978); therefore, data on just one of the two components of the soil biota can be misleading. For this reason, we emphasize the need to incorporate studies of the total soil biota in any program designed to assess and/or monitor reclamation success.

Conclusions. Two conclusions come from our studies at Mine C regarding the poor-quality spoil materials derived from the overburden shales and sandstones in coal-bearing regions of the arid Southwest. First, the incorporation into spoil of an organic substance may (1) enhance physical characteristics such as the spoil's capacity to absorb and store water, (2) stabilize the surface against erosion, and (3) provide an initial

Table 8.2. Average numbers of soil microarthropods and nematodes extracted from barley straw remaining in litterbags buried at 10-cm depth in amended mine spoils and unmined soils.

Taxa	July 28, 1980				September 3, 1980				October 24, 1983			
	Unmined	Bark	Straw	Topsoil	Unmined	Bark	Straw	Topsoil	Unmined	Bark	Straw	Topsoil
ACARI												
Prostigmata	37	236	311	21	2426	389	884	34	9534	916	470	560
Cryptostigmata	0	11	0	0	34	52	0	0	5	5	1	20
Astigmata	0	0	0	0	6	21	0	0	12	19	13	0
Mesostigmata	1	5	6	0	49	107	39	0	19	40	17	1
INSECTA												
Collembola	8	120	0	0	22	53	14	149	12	294	262	24
PSOCOPTERA	4	36	16	1	5	17	19	0	0	0	0	0
NEMATODA	258	468	147	235	1404	2783	2161	521	2184	1938	1048	736

180

carbon and nutrient pulse to aid in the establishment of a vigorous soil biota and thereby accelerate soil formation. Our studies showed that organic amendments, in association with raw spoil overburden, can establish vegetation and soil biota with equal or greater success than has been achieved with traditional topsoiling/mulching procedures. Topsoiling may not be justifiable as a required spoil dressing in situations where sufficient pre-existing topsoils do not exist, and should therefore be considered on a case-by-case basis. Organic amendments are a promising and economical alternative to current reclamation practices in the Southwest.

A second conclusion is that the assessment of reclamation success should not be based soley upon the amount of vegetation which is established over the relatively short bonding periods imposed by most regulatory agencies (see Chapter 9 for a discussion of the present regulations). Monitoring programs which attempt to integrate producer and decomposer system activities are best for assessing the long-term success of arid land reclamation.

Other studies of soil biology and reclamation

Our study demonstrates the potential value of existing reclaimed surfaces and experimental plots for the study of soil ecology and reclamation. Other researchers have also used existing reclaimed areas in the San Juan Basin for their studies of various soil biological functions. Miller et al. (1979) measured microbial populations on mine spoils ranging from one to fourteen years in age for valuable baseline data on the microbial dynamics in southwestern mine spoils. Fresquez and Lindemann (1982) compared microbial numbers, dehydrogenase activity, and fungal genera on barren spoil, stockpiled topsoil, reclaimed topsoiled spoil, and native soil at Mine C. They concluded that providing an available carbon source was more effective than providing an inoculant source in order to achieve a diverse, active spoil microflora.

The relative lack of studies regarding soil organisms and their various potential roles in mined-land reclamation is not peculiar to the San Juan Basin, or even to the western U.S. However, interest in the subject is on the rise. Parkinson (1978) provided a history of mine-related microbial and mycorrhizal studies. A good review of microbial studies associated with surface-mining activities in the western United States is given by Cundell (1977).

Studies which have addressed microarthropod activity associated with surface-mine reclamation are even rarer. There are only a few studies at mines around the United States of the invasion of spoil materials by

182 Walter G. Whitford and Ned Z. Elkins

different species of arthropods (Neumann 1978, Cross and Wilman 1982, Hawkins and Cross 1982).

Summary

There is little question that organic amendments will benefit reclamation in the arid Southwest. This chapter has shown that organic matter is important for the establishment of the soil biota, including mycorrhizae (Zak and Parkinson 1982, 1983). Adequate testing of the suitability of organic amendments requires careful examination of the entire soil biota from the systems or *holistic* perspective. Only when the proper mix of organic materials is applied to mined areas in arid regions can we expect to accelerate the processes of decomposition, mineralization, and ecological succession, as well as to induce colonization by important soil-builders like subterranean termites. The final result of these reclamation measures will be an equilibrium between the biotic components and mineral processes that signals a well-functioning reclaimed ecosystem.

Acknowledgements

The work on soil ecology and mine reclamation would not have been possible without the support of grants from the NSF (DEB 7716633, DEB 8020083 and BSR 821539) and from NMEI (2-68-3212). The manuscript was reviewed by Dr. John Zak, who added to the section on mycorrhizae and drew the mycorrhizal figure. Emma MacKay did the artwork.

References

Allen, M. F., W. K. Smith, T. S. Moore, Jr., and M. Christensen. 1981. Comparative water relations and photosynthesis of mycorrhizal and non-mycorrhizal *Bouteloua gracilis* H. B. K. Lag ex Steud. New Phytologist 88:683-693.

Anderson, R. V., E. T. Elliot, T. P. McClell, D. C. Coleman, C. V. Cole, and H. Hunt. 1978. Trophic interactions in microorganisms as they effect energy and nutrient dynamics. III. Biotic interactions of bacteria, amoeba, and nematodes. Microbial Ecology 4:361-371.

Anderson, R. V., D. C. Coleman, C. V. Cole, E. T. Elliot, and J. F. McClellan. 1979. The use of microcosms in evaluating bacteriophagic nematode responses to other organisms and effects on nutrient cycling. International Journal of Environmental Studies 13:175-182.

Babuik L. A., and E. H. Paul. 1970. The use of fluorescein isothiocyanate in the determination of the bacterial biomass in a grassland soil. Canadian Journal of Microbiology 16:57-62.

Cole, C. V., E. T. Elliot, H. W. Hunt, and D. C. Coleman. 1978. Trophic interactions in soils as they affect energy and nutrient dynamics. VI. Phosphorus transformations. Microbial Ecology 4:381-387.

Coleman, D. C., C. V. Cole, H. N. Hunt, and D. A. Klein. 1978. Trophic interactions in soil as they affect energy and nutrient dynamics. I. Introduction. Microbial Ecology 4:345-349.

Cross, E. A., and J. M. Wilman. 1982. A preliminary study of the refaunation of alkaline shale coal surface mine spoil by soil microarthropods. 1982 Symposium on Surface Mining Hydrology, Sedimentology, and Reclamation, December 5-10.

Crossley, D. A., Jr. 1977. The role of terrestrial saprophagous arthropods in forest soils: current status of concepts. *In* W. J. Mattson, ed., The role of arthropods in forest ecosystems, 49-56. New York, NY: Springer-Verlag.

Cundell, M. 1977. The role of microorganisms in the revegetation of strip-mined land in the western United States. Journal of Range Management 30(4):299-305.

Cutler, D. W. 1970. A method for estimating the number of active protozoa in the soil. Journal of Agricultural Science 10:135-143.

Durrell, L. W., and L. M. Shields. 1960. Fungi isolated in culture from soils of the Nevada test site. Mycologia 52:636-641.

Edwards, C. A., D. E. Reichle, and D. A. Crossley, Jr. 1970. The role of soil invertebrates in turnover of organic matter and nutrients. *In* D. E. Reichle, ed., Ecological studies. Vol. 16, Analysis of temperate forest ecosystems, 142-172. New York, NY: Springer-Verlag.

Elkins N. Z. 1983. Potential mediation by desert subterranean termites in infiltration, run-off and erosional soil loss on a desert watershed. Dissertation. Las Cruces, NM: New Mexico State University.

Elkins, N. Z., and W. G. Whitford. 1982. The role of microarthropods and nematodes in decomposition in a semi-arid ecosystem. Oecologia 55:303-310.

Elkins, N. Z., L. W. Parker, E. Aldon, and W. G. Whitford. 1984. Responses of soil biota to organic amendments in stripmine spoils in northwestern New Mexico. Journal of Environmental Quality. In press.

Elliot, E. T., R. V. Anderson, D. C. Coleman, and C. V. Cole. 1980. Habitable pore space and microbial trophic interactions. Oikos 35:335-375.

Ettershank, G., N. Z. Elkins, P. F. Santos, W. G. Whitford, and E. F. Aldon. 1978a. The use of termites and other soil fauna to develop soils on strip mine spoils. U.S. Forest Service research note RM-361.

Ettershank, G., J. Ettershank, M. Bryant, and W. G. Whitford. 1978b. Effects of nitrogen fertilization on primary production in a Chihuahuan desert ecosystem. Journal of Arid Environments 1:135-139.

Floret, C., R. Pontanier, and S. Rambal. 1982. Measurement and modelling of primary production and water use in a south Tunisian steppe. Journal of Arid Environments 5:77-90.

Franco, P. J., E. V. Edney, and J. F. McBrayer. 1979. The distribution and abundance of soil arthropods in the Northern Mojave desert. Journal of Arid Environments 2:137-150.

Freckman, D. W. 1982. Parameters of the nematode contribution to ecosystems. *In* D. W. Freckman, ed., Nematodes in soil ecosystems, 87-91. Austin, TX: University of Texas Press.

Freckman, D. W., and R. Mankau. 1977. Distribution and trophic structure of nematodes in desert soils. Ecology Bulletin (Stockholm) 25:511-514.

Freckman, D. W., R. Mankau, and H. Ferris. 1975. Nematode community structure in desert soils: Nematode recovery. Journal of Nematology 7:343-346.

Freckman D. W., and C. Womersley. 1983. Physiological adaptation of nematodes in Chihuahuan desert soils. *In* P. N. LeBrun, H. M. Andre, A. DeMedts, C. Gregoire-Wibo, and G. Wauthy, eds., Proceedings of the 8th international colloquium of soil zoology, 395-405. Belgium: Dieu-Brichart, O'Higmes Louvain-La-Neuve.

Fresquez, P. R., and W. C. Lindemann. 1982. Soil and rhizosphere microorganisms in amended coal mine spoils. Soil Science Society of America Journal 46:751-755.

Goodell, P. B., and H. Ferris. 1980. Plant parasitic nematode distributions in an alfalfa field. Journal of Nematology 12:136-140.

Greenslade, P. 1981. Survival of collembola in arid environments: observations in South Australia and the Sudan. Journal of Arid Environments 4:219-228.

Griffin, D. M. 1972. Ecology of soil fungi. London: Chapman and Hall.

Griffin, D. M. 1981. Water and microbial stress. Advances in Microbial Ecology 5:91-135.

Hanlon, R. D. G., and J. M. Anderson. 1979. The effects of collembola grazing on microbial activity in decomposing leaf litter. Oecologia 38:93-99.

Haverty, M. I., J. P. LaFage, and W. L. Nutting. 1974. Seasonal activity of environmental control of foraging of the subterranean termite, *Heterotermes aureus* (Snyder) in a desert grassland. Life Sciences 15:1091-1101.

Hawkins, B. A., and E. A. Cross. 1982. Patterns of refaunation of reclaimed strip mine spoils by nonterricolous arthropods. Environmental Entomology 11:3.

Howard, R. W., and M. I. Haverty. 1980. Reproduction in mature colonies of *Reticalitermes flavipes*: Abundance, sex-ratio, and association with soldiers. Environmental Entomology 9:458-460.

Jenkinson, D. S., and D. S. Powlsen. 1976. The effects of biocidal treatment on metabolism in soil. V. A method for measuring soil biomass. Soil Biology and Biochemistry 8:209-213.

Jones, P. C. T., and J. E Mollison. 1948. A technique for quantitative estimation of soil microorganisms. Journal of General Microbiology 2:54-69.

Khan, A. G. 1981. Growth response of endomycorrhizal onions in unsterilized coal wastes. New Phytolologist 87:363-370.

Krantz, G. W. 1978. A manual of acarology. 2nd edition. Corvallis, OR: Oregon State University Bookstores, Inc.

Lambert, D. H., and H. Cole, Jr. 1980. Effects of mycorrhizae on establishment and performance of forage species in mine spoils. Agronomy Journal 72:257-260.

Lee, K. E., and T. G. Wood. 1971a. Termites and soil. London and New York: Academic Press.

Lee, K. E., and T. G. Wood. 1971b. Physical and chemical effects on soils of some Australian termites and their pedological significance. Pedobiologia 11:376-409.

Marx, D. H. 1975. Mycorrhizae and the establishment of trees on strip-mined land. Ohio Journal of Science 75:288-297.

Maser, C., J. M. Trappe, and R. A. Nussbaum. 1978. Fungal-small mammal interrelationships with emphasis on Oregon coniferous forests. Ecology 59:799-809.

Mellilo, J. M., J. D. Aber, and J. F. Muratore. 1982. Nitrogen and lignin control of hardwood leaf litter decomposition dynamics. Ecology 63:621-626.

Miller, R. V., E. E. Staffeldt, and B. C. Williams. 1979. Microbial populations in undisturbed soils and coal mine spoils in semi-arid conditions. U.S. Forest Service research note RM-372.

Neumann, U. 1978. Succession of soil fauna in afforested spoil banks of the Brown Coal Mining district of Cologne. *In* M. K. Wali, ed., Ecology and Coal Resource Development, 335-348. New York, NY: Pergamon Press.

Nicholas, W. L. 1975. The biology of free-living nematodes. Oxford: Clarendon Press.

Nicholas, W. L., and S. Viswanathan. 1975. A study of nutrition of *Caenothabditis briggsae* (Rhabditidae) fed on ^{14}C and ^{32}P labeled bacteria. Nematologia 21:385-400.

Page, A. L., R. H. Miller, and D. R. Keeney, eds. 1982. Methods of soil analysis, chemical and microbiological properties. 2nd edition. Madison, WI: American Society of Agronomy and Soil Science Society of America.

Parker, L. W., H. G. Fowler, G. Ettershank, and W. G. Whitford. 1982. The effects of subterranean termite removal on desert soil nitrogen and ephemeral flora. Journal of Arid Environments 5:53-59.

Parker, L. W., P. F. Santos, J. Phillips, and W. G. Whitford. 1984a. Carbon and nitrogen dynamics during the decomposition of litter and roots of a Chihuahuan desert annual, *Lepidium lasiocarpum*. Ecological Monographs. In press.

Parker, L. W., N. Z. Elkins, E. F. Aldon, and W. G. Whitford. 1984b. Development of soil biota and nutrient cycles on reclaimed coal mine spoils in the arid Southwest. Journal of Environmental Quality. In press.

Parkinson, D. 1978. Microbes, mycorrhizae and mine spoil. *In* M. K. Wali, ed., Ecology and coal resource development, 634-642. New York, NY: Pergamon Press.

Parkinson, D., S. Visser, R. M. Danielson, and J. Zak. 1980. Restoration of fungal activity in tailing sand (Oil Sand). *In* D. Dindall, ed., Soil biology as related to land use practices, 362-370. Washington, DC: UNESCO.

Penning de Vries, F. W. T., and M. A. Djiteye, eds. 1982. La productivite des paturages saheliens. Wageringu: Centre for Agriculture Publishing and Documentation.

Phillipson, J. 1971. Methods of study in quantitative soil ecology: population production and energy flow. Oxford: Blackwell Scientific Publications.

Reeves, F. B., D. Wagner, T. Moorman, and J. Kiel. 1979. The role of endomycorrhizae in revegetation practices in the semi-arid west. I. A comparison of incidence of mycorrhizae in severely disturbed vs. natural environments. American Journal of Botany 66:6-13.

Santos, P. F., E. DePree, and W. G. Whitford. 1978. Spatial distribution of litter and microarthropods in a Chihuahuan desert ecosystem. Journal of Arid Environments 1:41-48.

Santos, P. F., and W. G. Whitford. 1981. The effects of microarthropods on litter decomposition in a Chihuahuan desert ecosystem. Ecology 62:654-663.

Santos, P. F., J. Phillips, and W. G. Whitford. 1981. The role of mites and nematodes in early stages of buried litter decomposition in a desert. Ecology 62:654-663.

Santos, P. F., N. Z. Elkins, Y. Steinberger, and W. G. Whitford. 1984. A comparison of surface and buried *Larrea tridentata* litter decomposition in North American hot deserts. Ecology 65:278-284.

Schaefer, D. A., and W. G. Whitford. 1981. Nutrient cycling by the subterranean termite *Gnathamitermes tubiformans* in a Chihuahuan desert ecosystem. Oecologia 48:277-283.

Schoknecht, J. D., and M. J. Hattingh. 1976. X-ray microanalysis of elements in cells of VA mycorrhizal and nonmycorrhizal onions. Mycologia 68:296-303.

Scholl, D. C., and C. P. Pase. 1984. Response of certain wheatgrasses to organic amendments and contour furrowing on a New Mexico coal mine spoil. Journal of Environmental Quality 13:479-482.

Singh, B. N. 1946. A method of estimating the numbers of soil protozoa, especially amoebae, based on their differential feeding on bacteria. Annals of Applied Biology 33:112-119.

Swift, M. J., O. W. Heal, and J. M. Anderson. 1979. Decomposition in terrestrial ecosystems. Los Angeles, CA: University of California Press.

Taylor, E. C. 1979. Seasonal distribution and abundance of fungi in two desert grassland communities. Journal of Arid Environments 2:195-312.

Trappe, J. M. 1981. Mycorrhizae and productivity of arid and semi-arid rangelands. *In* Advances in food producing systems for arid and semi-arid lands, 581-599. New York, NY: Academic Press.

Van Veen, J. H., and E. A. Paul. 1979. Conversion of biovolume measurements of soil organisms, grown under various moisture tensions, to biomass and their nutrient content. Applied Environmental Microbiology 37:606-692.

West, N. E., and J. O. Klemmedson. 1978. Structural distribution of nitrogen in desert ecosystems. *In* N. E. West, and J. J. Skujins, eds., Nitrogen in desert ecosystems, 1-16. Strandsburg, PA: Hutchinson and Ross Inc.

Whitford, W. G., D. W. Freckman, N. Z. Elkins, L. W. Parker, R. Parmalee, J. Phillips, and S. Tucker. 1981a. Diurnal migration and responses to simulated rainfall in desert soil microarthropods and nematodes. Soil Biology and Biochemistry 13:417-425.

Whitford, W. G., V. Meentemeyer, T. R. Seastedt, K. Cromack, Jr., D. A. Crossley,

P. Santos, R. L. Todd, and J. B. Waide. 1981b. Exceptions to the AET model: deserts and clear-cut forests. Ecology 62:275-277.

Whitford, W. G., D. W. Freckman, P. F. Santos, N. Z. Elkins, and L. W. Parker. 1982a. The role of nematodes in decomposition in desert ecosystems. *In* D. W. Freckman, ed., Nematodes in soil ecosystems, 98-116. Austin, TX: University of Texas Press.

Whitford, W. G., Y. Steinberger, and G. Ettershank. 1982b. Contributions of subterranean termites to the "economy" of Chihuahuan desert ecosystems. Oecologia 55:298-302.

Wiesner, F. M. 1965. The termites of the United States. Elizabeth, NY: The National Pest Control Association.

Witkamp, M., and D. A. Crossley. 1966. The role of arthropods and microflora in breakdown of white oak litter. Pedobiologia 6:293-303.

Zak, J. C., and D. Parkinson. 1982. Initial vesicular-arbuscular mycorrhizal development of slender wheatgrass on two amended mine spoils. Canadian Journal of Botany 60:2241-2248.

Zak, J. C., and D. Parkinson. 1983. Effects of surface amendment of two mine spoils in Alberta, Canada, on vesicular-arbuscular mycorrhizal development of slender wheatgrass: a 4-year study. Canadian Journal of Botany 61:798-803.

9

Vegetation Ecology, Sample Adequacy, and the Determination of Reclamation Success

N. Timothy Fischer

Introduction

The goal of reclamation is to convert land devastated by mining to a condition which is stable, nonpolluting, and of equal or greater value than the pre-mining condition. Revegetation plays a vital role in the attainment of this goal. A good vegetative cover on the reclaimed surface reduces both wind and water erosion, thereby aiding in the stabilization of the surface and reducing the release of sediments to outwash and of dust to the atmosphere. More importantly, vegetation provides a resource base for certain post-mining land uses, such as grazing, wood production, and wildlife.

Adequate revegetation is, therefore, an essential part of the legal requirements for reclamation. A failure by the mining company to meet the legal standards for revegetation can mean the loss of the *performance bond*, money posted by the company prior to mining and held by the government against the possibility of inadequate reclamation. Thus, the establishment of realistic and objective standards for revegetation is of major concern to both the mining companies and the regulatory agencies who are responsible for enforcing reclamation laws.

One approach to the problem of establishing revegetation success standards is to use the pre-mining vegetation as a model, against which the post-mining vegetation is evaluated. Since the natural vegetation at a site has developed under the local climate of the site, it is a good indicator of the potential of the site to support vegetation. This approach is widely used in the arid Southwest, where climatic conditions are characteristically erratic and highly variable from site to site.

This chapter examines the use of pre-mining vegetation as the standard for revegetation success. In addition to describing the current legal requirements for revegetation success and the basic techniques used in the quantitative characterization of vegetation, this chapter examines the conflict between the accuracy and efficiency (in terms of time and effort) of these techniques. This conflict has long been a source of controversy between the regulatory agencies, which must ensure that vegetation data are accurate and defensible, and the mining companies, which are concerned about cost of data collection.

Finally, this chapter discusses a more fundamental question: Do the currently accepted criteria for revegetation success truly indicate the establishment of a vegetative community which will continue to satisfy the objectives of reclamation for decades beyond the end of the mine's period of legal responsibility?

Current Standards for Revegetation Success

The current regulations for surface-mine reclamation require each operator to establish on all affected lands a vegetative cover which is "diverse, effective, and permanent." In addition, the vegetation must be capable of supporting the prescribed post-mining land use, which must be equivalent to or better than the pre-mining land use. The examples in this chapter address a situation where grazing, that is, the provision of rangeland, is the designated post-mining land use, as is the case for most mines in arid and semiarid habitats; however, the principles and problems involved in the comparison of pre- and post-mining vegetation communities are broadly applicable.

The fulfillment of revegetation requirements in arid habitats is presently evaluated in terms of two vegetation parameters: cover and productivity.

Vegetative cover is the proportion (expressed as a percentage) of the ground surface with plant material (stems, leaves, flowers, and so forth) directly above it. It can be thought of as the amount of ground surface shaded by plants when the sun is directly overhead. Thus, this parameter also indicates the amount of ground surface not exposed to the erosive forces of raindrop impact and wind. High coverage values indicate more effective prevention of erosion by vegetation.

Productivity is the amount of new plant material which is annually added to an area by the growth of plants. This parameter is expressed as the increase in plant biomass (weight) per unit area of land, that is, lbs./acre, or g/m^2. Productivity is a measure of the land's ability to support the post-mining land use.

In the San Juan Basin, the revegetated surface must go through ten years without additional seeding, fertilization, or irrigation before it can be tested for success. The demonstration of success requires that the cover and productivity of the revegetated surface be equal or greater than the predetermined standards for each growing season (spring and summer); the mining company must demonstrate this with 90% statistical confidence (80% for shrublands). If the standards are based upon the pre-mining vegetation, then the cover and productivity of the vegetation at the mine site must be sampled before mining begins.

The Analysis of Pre-mining Vegetation

The analysis of vegetation involves two general steps: (1) the division of the study area into vegetation types, and (2) the characterization of the vegetation within each type. The first step would be unnecessary if the vegetation were uniform across the landscape, but this is rarely the case. Instead, the natural vegetation in an area consists of different zones with different characterstics such as abundance, species composition, and general appearance. The identification and delineation of these vegetation zones is necessary to make the data obtained by vegetation sampling more manageable and meaningful. However, this process of identifying and classifying vegetation types is difficult and controversial, and will be discussed at some length before dealing with the actual characterization of vegetation within each type.

Vegetation classification

Classification is a tool of scientific inquiry which provides a hierarchical ordering of objects under investigation. This ordering may reflect suspected natural relationships between the objects, such as the classification of species according to evolutionary relationships, or it may be artificially constructed to facilitate the understanding of a system, such as the classification of plants according to their *growth form*, that is, whether they are grasses, forbs, shrubs, or trees.

The study of natural vegetation is greatly facilitated by its subdivision into relatively homogeneous units, or *vegetation types.* Many systems have been devised for defining and classifying vegetation types; some reflect particular theories of vegetation processes and dynamics, while others are more functionally oriented, that is, applicable to resource inventories or management plans. The plurality of systems can be the source of great confusion to novices in vegetation research. Misuse of these classification systems can be misleading.

This section briefly describes two basic approaches to vegetation classification, both of which attempt to make divisions into natural units.

One approach is based upon sociological relationships among plants; the other is based on habitat (environmental) characteristics. For more in-depth reviews of this topic, see Daubenmire (1968), Mueller-Dombois and Ellenberg (1974), Oosting (1956), and Whittaker (1962 and 1978).

Sociological classification.   The plants which comprise natural vegetation are seldom scattered haphazardly over the landscape, but instead tend to be segregated into repeating groups of species, which are distinct from neighboring groups. Early workers in the field of *vegetation ecology*, which is the study of the structure and function of natural vegetation (Mueller-Dombois and Ellenberg 1974), were intrigued by patterns in natural vegetation. Researchers attempted to explain these patterns in terms of the interactions between plants or plant species. The study of interrelationships of plants is termed *plant sociology* (also phytosociology or plant synecology).

Sociological classification systems attempt to divide the natural vegetation into units which approximate the sociological structure of the vegetation. The theoretical unit of sociological structure is the *community*, defined as "an aggregation of living organisms having mutual relationships among themselves and to their environment" (Oosting 1956). A *plant community* refers to a community in which only the vegetational composition is considered.

Since the actual interrelationships between plants cannot be observed directly, plant communities cannot be physically defined. They are, in essence, theoretical entities which are represented in sociological classification systems by *associations*. An association was defined by the Third International Botanical Conference in 1910 as "a plant community of definite floristic (species) composition, presenting a uniform physiognomy (i.e., forest, woodland, grassland, etc.), and growing in uniform habitat conditions" (Komarkova 1983). Note that the species interactions are only assumed in this definition. The actual identification of the association is based strictly upon the species composition and the environment.

Higher units of classification can be formed by grouping similar associations, then grouping these groups, and so forth. Lower units can be formed by successively subdividing the association.

A recent example of a vegetation classification system which is basically sociological is the digitized system proposed by Brown, Lowe, and Pase (1979). This system is designed to accommodate all levels of organization, from the vegetation of continents to very local communities. The eight levels used in the system are: the biogeographic realm, the vegetation, the formation type (or biome type), the climatic zone, the major biotic community (or biome), the climax series, the association, and the phase.

Digitization of this system involves the assignment of a single digit to each level and ordering these digits for the eight levels in an eight-place decimal of the form 1,234.5678, the first digit denoting the biogeographic realm, the second the vegetation (upland or wetland), and so on. For example, much of the San Juan Basin is classified as the Grama "short-grass" Series and is assigned the number 142.12: 1 = Upland Vegetation, 4 = Grassland Formation, 2 = Cold-Temperate Climate, 1 = Plains Grassland, and 2 = the Grama "short-grass" Series (the Biogeographic Realm [Nearctic] is not specified in the number). This climax series is composed of several associations defined according to their dominant species. The digitized approach allows vegetation data to be handled easily by computers; if used consistently, this approach greatly facilitates resource inventories.

The large number of classification systems proposed over the years reflects the diversity of the ways in which ecologists have regarded vegetation structure. Initially, ecologists considered communities to be discrete, mappable units, within which species contributed to each other's survival; the boundaries between communities were areas of conflict, as each community tried to expand into the other's territory (Warming 1909). This idea of "closed communities" gradually fell out of favor, with the result that boundaries between "communities" became fuzzier. Finally, in the 1950s and 1960s, the community concept was replaced by the *continuum concept*, developed primarily by Whittaker (1956, 1962) and McIntosh (1967) using the ideas of Gleason (1939).

In the continuum concept, the environment is seen as consisting of gradients, that is, gradual spatial changes in environmental conditions. The vegetation gradually changed along these gradients, with vegetation types grading into each other; boundaries *per se* did not exist. Every location had its own unique vegetation, thus denying the presence of repetitious units or patterns. This interpretation did not lend itself to the traditional approaches of vegetation classification.

It is generally agreed that the continuum concept provides the most accurate model of vegetation structure; however, this model limits the manageability and utility of vegetation data. This concept calls for methods of *gradient analysis* which require a computer and which produce results that are of limited management value because they cannot be generalized to equivalent units at other locations. For inventory and management applications, the classification of vegetation into discrete, homogeneous, and mappable units must still take precedence over the accurate analysis of community structure.

Habitat classification. The classification of vegetation based on suspected community relationships has many inherent problems. Communities are unbounded in space and time. Any attempt to draw a geographical boundary around a community must be accompanied by the recognition that the boundary is an artificial construct, which does not and cannot truly represent a sharp line of transition between separate groups of interacting organisms.

There is another approach to vegetation classification which does allow a geographic definition of the units, which are called *habitat types*. Daubenmire (1968) describes habitat as a

> constellation of interacting physical and biological factors which provide at least minimal conditions for one organism to live or for a group to live together.

A habitat type is the set of all areas with equivalent habitats. Note that it is the environment which is the basis of classification, not the community.

It has been said that the plant species on an area are symptoms of an ecological syndrome, that is, plants can be used as indicators of environmental conditions. Plants which live together on a site must share a common tolerance to the conditions of the site. Thus, recurring combinations of species must indicate either identical environmental conditions, or different environments which have an equivalent effect on vegetation, that is, the habitat type.

Habitat types are identified by the plant species which comprise them; thus, they are approximately equivalent to associations. However, habitat types differ in that they represent a range of environmental conditions, not an assemblage of interacting species. The power of this difference can be seen when the habitat approach is applied to an environmental gradient where two communities blend into each other. Although the approaches define similar vegetation types, the boundary between the types is still arbitrary if drawn on sociological grounds, because the two communities are broadly interconnected. However, when the gradient is analyzed in terms of habitats, the decision is made to allow a certain amount of intrahabitat variability, and the gradient is divided accordingly, with the divisions representing the range of conditions within the gradient (Fig. 9.1).

Two problems exist with the use of habitat types. First, differences in vegetational composition between areas are not necessarily due to differences in environment. Disturbances such as fire, drought, and overgrazing can result in differences in vegetation within a single habitat type, which can be misinterpreted as resulting from differences in habitat. Second, habitat types are defined as *equivalent* environments, not as equal environments. This presents problems when the system is used for land-management decisions, because different environments which produce the same vegetation will not necessarily respond the same way to either natural or man-caused disturbances, including management practices.

Range managers have long recognized these shortcomings of the habitat type and have modified it to a system based on *range sites*. Range sites differ from habitat types primarily by the fact that they are defined according to physical and biotic factors of the environment, and not just according to vegetation (Anderson 1983). Thus, the range site is a product of the climate, topography, soil, and biota of the site. Since the soil itself is a product of the climate, topography, geology, and biota over time (Brady 1974), the range site and soil type should be approximately equivalent in geographical extent. In fact, range site boundaries are often made directly from soil maps.

Finally, where greater detail is desired, a range site might be further subdivided, usually by topographic criteria such as slope and orientation, into *ecological response units* (Bonham, et al. 1980). This level of consideration is particularly well suited to inventories involved in surface mining. These units are highly homogeneous with respect to environmental factors; therefore, they produce a high degree of confidence that similar units will respond equally to management.

Vegetation sampling methods

The measurements of cover and productivity are difficult and time consuming. It is essentially impossible to measure these two parameters directly on a large area. Instead, *samples* are taken of the vegetation. Samples are subsets of any measured system which are intended to represent the entire system. Since it is impossible to measure every plant on a large area such as a mine site, vegetation sampling is the only way to obtain data for the evaluation of vegetation before mining and after reclamation.

The basic principle of sampling is that a part is representative of the whole; therefore, a measurement of the part (a sample) is an estimate of the measurement of the whole. Thus, in the case of vegetation, if one can measure the cover or productivity of a portion of the area, that value can be used as an estimate of that parameter for the entire area. These measurements are usually made within a *stand*, a defined area of uniform vegetation chosen to represent a unit of vegetation classification. Some of the basic methods of vegetation sampling are described below.

Quadrats.   Quadrat generally refers to any two-dimensional area within which vegetation is measured. As the name implies, a quadrat is usually square or rectangular in shape; however, circular quadrats are also

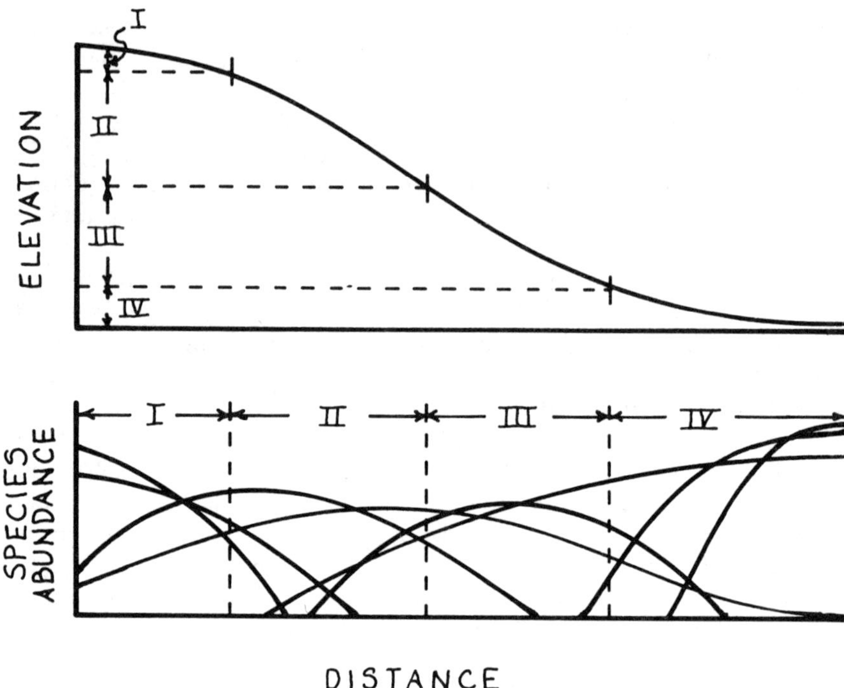

Figure 9.1. A continuum of vegetation change (lower graph) along a hillslope (upper graph) which has been arbitrarily divided into four habitat types, each of which occupies a discrete elevational range.

commonly used. The size of the quadrat varies widely, but should be chosen carefully according to the needs of the study. Measurements of forest trees, for example, will require larger quadrats than those of grasses and forbs. Productivity is best measured by clipping and weighing plants from within a quadrat.

    Lines. Cover can only be roughly estimated in a quadrat. A more accurate approach is the line-intercept method, wherein a line (measuring tape) is laid across the vegetation, and the proportion of the line which is intercepted by a vertical projection of plant material serves as a precise estimate of the stand cover.

    Points. Another approach to measuring cover involves probability theory. If the coverage of the stand is 30%, the probability that a randomly located point within the stand will strike a plant is also 30%. Thus, by locating several such points within the stand, the proportion of those striking plants will approximate the stand coverage value. Many variations of this method are used, including arranging the points in lines or grids.

"Plotless methods." A variety of sampling methods are considered plotless, in that the sampling units have no boundaries. These methods examine characteristics of the plant community, such as the distance between individuals or the distance from a random point to the nearest plant. These methods have been devised for quick measurements of cover and density and are particularly useful for large plants, primarily tree species. These include the Bitterlich variable-radius method, the point-centered-quarter method, and the nearest-neighbor method (Mueller-Dombois and Ellenberg 1974).

Sample Adequacy

If a stand could be defined which was completely uniform, any single sample would accurately estimate the entire stand. Unfortunately, such uniformity is never attained in natural vegetation. A certain degree of variability in the stand must always be allowed. Thus, the stand parameters are best estimated by the average value of several samples. The more samples taken, the better the estimate will be. But what is the minimum number of samples to obtain a good estimate?

The test for adequate number of samples (or sample size) is one of the most contested issues in surface-mine vegetation analysis. This section reviews some of the methods which have been proposed and those which are currently in use. These methods fall into two basic categories: empirical and statistical. These will be treated separately, although they are related in ways that will be discussed.

Empirical methods

The empirical methods for determining adequate sample size are those which are based ultimately on a value judgement by the experimenter, instead of statistical assessment (although empirical methods may contain elements of statistics). The advantages of this approach are: (1) the freedom from assumptions about the nature of the data, which must be made before statistical methods can be validly applied; and (2) the smaller sample sizes often dictated by these tests. The greatest disadvantage to the empirical approach is the inability to attach precise, defensible confidence levels to the resulting estimations.

Maximum sample size.   The simplest test for sample adequacy is to choose arbitrarily a maximum sample size prior to sampling and quit when this number has been reached. Although this method is extremely crude, allowing no possible estimation of accuracy, it is used in several western states when other methods dictate seemingly excessive sample sizes. These maximum values are, of course, not completely arbitrary, but are based on experience in vegetation sampling. However, considering the range of values used by the various states, from forty to sixty (Keammerer 1983), it is apparent that experience has taught different people different things.

"Behavioral methods."   The other empirical methods can be classed as "behavioral," since they depend upon the behavior of a particular parameter as samples are taken. The behavior is analyzed either with graphs of the parameter plotted against sample size, or with numerical criteria, such as the percentage of change in the parameter at different points in the acquisition of samples.

The parameter chosen for this type of analysis must behave in a predictable manner with increasing sample size; for instance, the value for the parameter may approach some maximum value, a mean value, or zero. Parameters which have been used include: the number of species in the entire sample, the sample mean, the sample variance, the coefficient of variation, and the ratio between the standard error of the mean and the mean. Each of these will be defined later.

These tests require an *a priori* establishment of success criteria, such as a maximum range of variation in the parameter over a certain number of consecutive samples, or the value of the parameter falling below a given value. When these criteria are strict, much confidence can be placed on the accuracy of the result. However, these criteria are purely subjective and must not be confused with the numerically defined confidence limits which are involved in the statistical methods.

(1) Species-area curve.   The species-area curve is one of the oldest methods of determining adequate sample size in vegetation ecology. In this method, the total number of species encountered is plotted against the sample size (or the total area sampled). When samples are taken within a single vegetation unit, the resulting curve will initially rise sharply as the most common species are encountered, but will level off as only rare species are added to the list (Fig. 9.2). The inflection point of this curve is usually the indicator of sample adequacy. It should be noted, however, that the purpose of this test is to determine how well the sample represents the composition and diversity of the species community (Pielou 1977). It does not necessarily indicate the effectiveness of the sample in representing vegetation parameters such as cover, productivity, density, and so forth.

(2) Running mean.   A more direct method of assessing vegetation

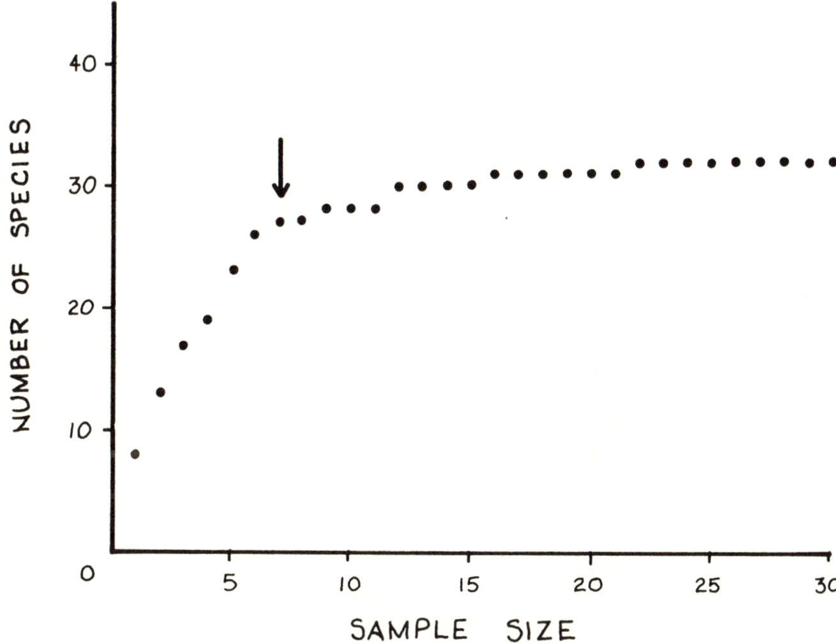

Figure 9.2.   A species-area curve with arrow indicating the inflection point characteristic of these curves.

parameters is through a graph of the running mean of the sample. The sample mean is a statistical parameter which, for our purposes, will simply be defined as the average value of a parameter for a set of samples. In a running mean, this average is recalculated each time a new sample is added and is then plotted against sample size (Fig. 9.3). Sowards (1983) proposed the use of the running mean as a test for sample adequacy in vegetation surveys on mine sites. In this method, the fluctuation of the mean over a sequence of ten samples is analyzed. If this fluctuation remains within set boundaries, the sample is adequate. More will be said about this later.

(3) Sample variance.   A similar approach is possible using another statistical parameter, the sample variance. Variance is a measure of the heterogeneity of the sample values (which will also estimate the heterogeneity of the stand). Variance is calculated by the following formula:

$$S^2 = \frac{\Sigma\,(x-\bar{X})^2}{N-1} \tag{1}$$

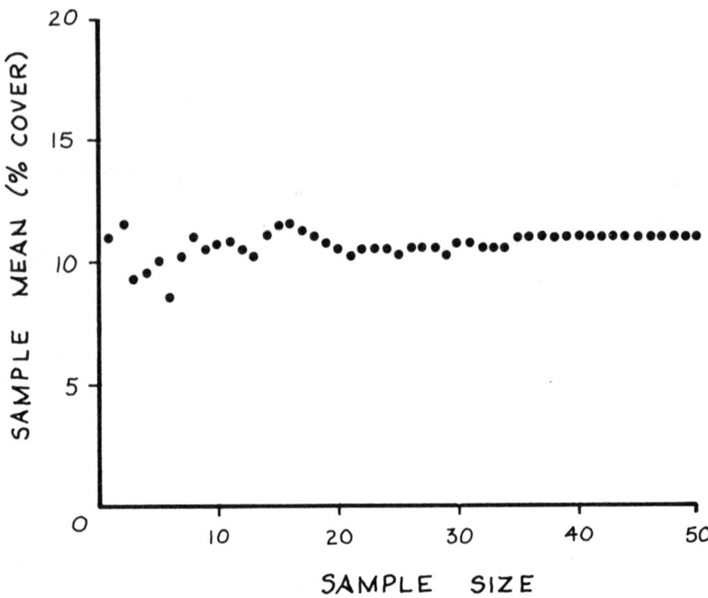

Figure 9.3.   Example of a running mean.

where S^2 is the variance, x is the value of a single sample, $\bar{X}$ is the sample mean, and N is the number of samples. The use of variance as a basis for sample adequacy is mentioned by Keammerer (1983), but has not otherwise received the attention it may deserve as a basis for testing sample adequacy.

(4) **Coefficient of variation.**   The square root of the variance is called the *standard deviation*. The standard deviation divided by the mean is the coefficient of variation (CV). Keammerer (1983) states that the CV will also level off when plotted against sample size. This is expected since, as the sample size increases, both the sample mean and sample variance are approaching their true values in the stand. However, it is questionable whether the use of the CV as the basis for determining adequacy has any advantage over the previously mentioned parameters.

(5) **SEM:mean ratio.**   A final example of the empirical method is the use of the ratio between the standard error of the mean (SEM, calculated by dividing the standard deviation by the square root of the sample size) and the mean. This ratio approaches zero as sample size increases

due to the decreasing value of the SEM. Adequacy is defined as the sample size for which this ratio falls below a given small value, A:

$$\frac{\frac{S}{\sqrt{N}}}{\bar{X}} < A \tag{2}$$

where S is the standard deviation, N is the sample size, and $\bar{X}$ is the mean.

Note that this inequality can be solved for N to give a direct calculation of the adequate sample size from the square of the CV:

$$\frac{\frac{S}{\sqrt{N}}}{\bar{X}} < A \qquad N > A^{-2}\left(\frac{S^2}{\bar{X}^2}\right) \qquad N > A^{-2}(CV)^2 \tag{3}$$

The paradox in this formula is that the reliability of the calculated value of N depends on the accuracy of the estimation of the CV, which in turn depends on N. Enough samples must be taken to ensure a good estimate of the CV before N can be calculated by this method. Thus, this test depends upon the prior fulfillment of the adequacy test for the coefficient of variation.

Statistical methods

Statistics is a branch of mathematics which deals with variables whose values are determined by probability and chance. These are called *random variables*. The measurement of cover from a single sample is an example of a random variable, because its value depends upon the amount of vegetation which happens to occur within the sample. If the sample is randomly located in the stand, that is, without regard to vegetation patterns, this amount is controlled by chance. Therefore, if certain criteria are met in the collection of vegetation data, then statistical theory can be applied to the data. This means that statistical methods not only predict the number of samples required for accurate estimates of stand parameters, but they also indicate the probability that the estimate is within a particular range of accuracy.

An essential part of statistical analysis is the *frequency distribution*. As illustrated in Fig. 9.4, a frequency distribution is made from a set of values of a random variable (for example, cover measurements from a set of transects) by graphing the number of observations in intervals which describe ranges in the possible values for the random variable. When sample data are shown in this manner, several pieces of information are immediately apparent. The uniformity of the stand, for exam-

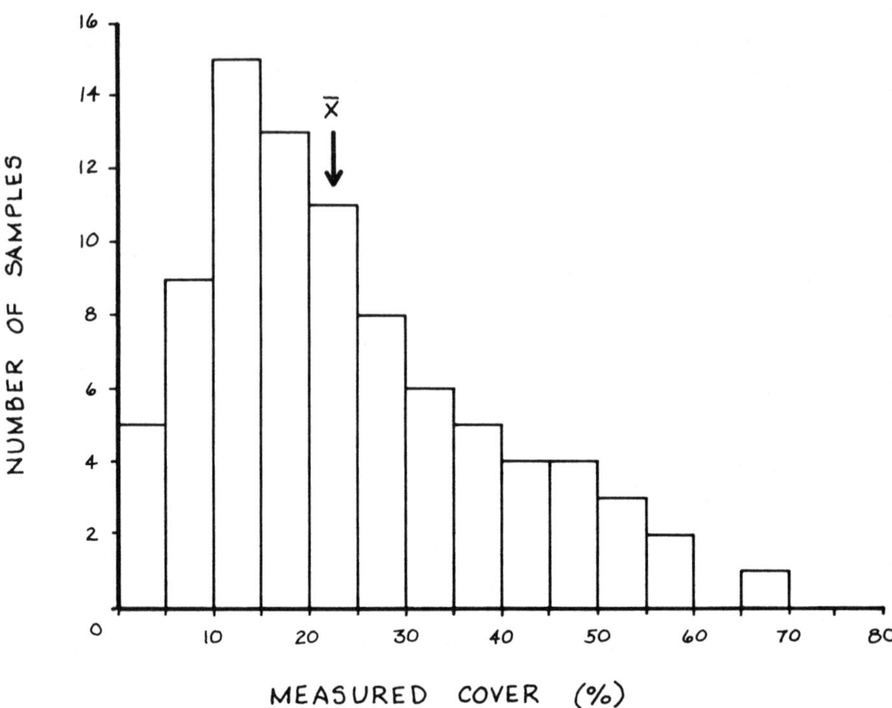

Figure 9.4. A frequency distribution of cover data exemplifying an asymmetrical distribution.

ple, can be observed. If the bars are concentrated in one small area, the stand is uniform; if they are scattered, the stand contains a high degree of variability. If there are two or more areas of concentration, there is a chance that more than one vegetation type has been included, and the stand delineation should be reconsidered.

Normally distributed data. Of particular importance to this discussion is the use of frequency distributions for determining the applicability of certain statistical tests. Several mathematically defined distributions have been devised which can be used as models for frequency distributions in nature. The most powerful of these is the *normal distribution*, often referred to as the bell-curve (Fig. 9.5). The normal distribution has a peak at the mean and is symmetrical around this point. Its spread is determined by the variance. When a normal distribution is well fitted to a data set and its mean and variance are determined, the probability that any particular sample will be within a certain distance from the mean can be determined by calculating Z, which is defined by the formula:

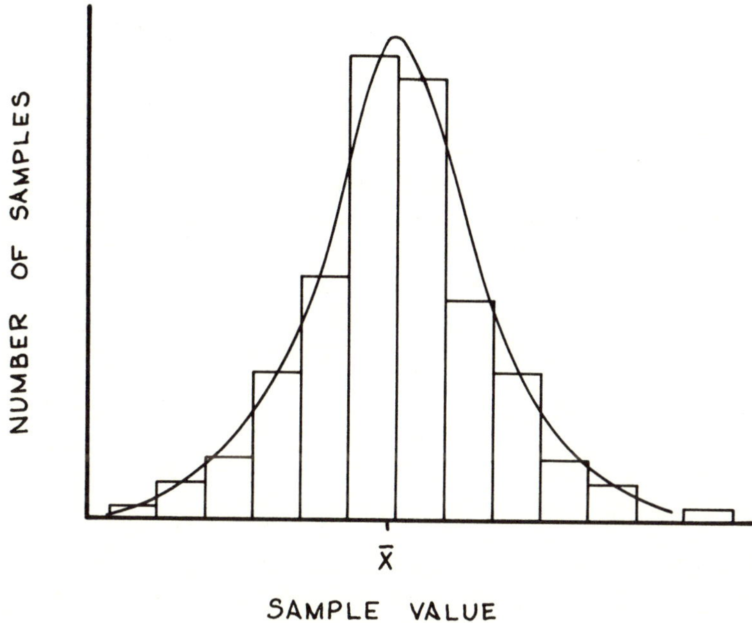

Figure 9.5. The normal distribution fitted to symmetrical frequency distribution.

$$Z = \frac{L}{S} \qquad (4)$$

where L is the distance from the mean and S is the standard deviation. The probability can be determined from a table of Z values developed for this purpose (and found in most statistics books).

Returning to the running-mean method described in the previous section, suppose that a data set of one hundred coverage values has a normal distribution and that a running mean was plotted while the samples were being taken. Suppose further that this plot indicated that the stand cover was accurately estimated after only thirty samples had been taken. This result is, in part, due to the order in which the one hundred samples were considered in the calculation. Another running mean computed from the same data set, but in a different order, might not satisfy the empirical test until thirty-five samples are considered. Another ordering might, by chance, satisfy the criteria after only twenty samples. By repeating this process many times and plotting the running means on

                  N. Timothy Fischer

the same graph, one would develop a distribution of points at each sample size: one for the samples taken singly, one for the average of two randomly selected samples, one for three samples, and so on (Fig. 9.6).

If the initial data set is normally distributed with a variance of S^2, it can be shown that each of the subsequent distributions (of means) will also be normally distributed with the same mean, but with a variance that decreases progressively as the sample size increases, according to the following relationship:

$$S_N^2 = \frac{S^2}{N} \tag{5}$$

where S_N^2 is the variance of the means of N randomly selected samples from a data set of variance S^2 (Mood et al. 1974). Note that the square root of this value is the SEM.

Since the means are normally distributed (assuming that the original data was normal) and the variance of the means can be calculated from the variance of the sample, then the probability that the actual stand value lies within a given distance (L) of the mean (of a sample of size N) can be determined from the Z-value:

$$Z = \frac{L}{\frac{S}{\sqrt{N}}} = \frac{L\sqrt{N}}{S} \tag{6}$$

This equation can be solved for N to get:

$$N = \frac{Z^2 S^2}{L^2} \tag{7}$$

(Snedecor and Cochran 1967)

This equation provides a method for calculating the number of samples (N) required to obtain a mean which is within L units of the true value (for the stand). This is done with a probability that is determined by the value of Z (a value called t is used instead of Z when the number of samples taken is less than thirty). When L is defined as a proportion of the mean, $L = p\bar{X}$, equation (7) can be written:

$$N = \frac{Z^2 S^2}{(p\bar{X})^2} = Z^2 p^{-2} \left(\frac{S^2}{\bar{X}^2}\right) = Z^2 p^{-2} (CV)^2 \tag{8}$$

which is very similar to equation (3) (identical when $A = p/Z$), but with statistical significance added to the constant term.

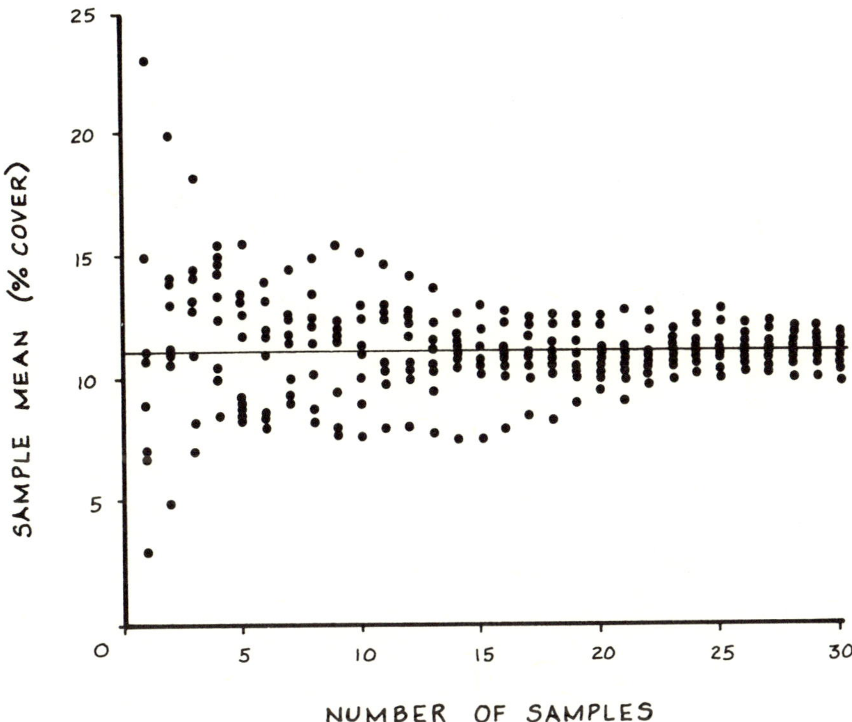

Figure 9.6. Eight running means from the same data set with an actual mean indicated at A.

The test in equation (7) is based on the requirement that the sample be normally distributed. Asymmetrical data can sometimes be mathematically transformed to achieve a distribution closer to normal (Bonham 1982, Sowards 1983), but the transformed data set must be used only for determining sample adequacy, not for calculating the stand parameters.

Deviations from normality can also be circumvented by the Central Limit Theorem, which states that the distribution of means always approaches the normal distribution as sample size increases (Mood et al. 1974). This is true for any initial frequency distribution. The rate at which it approaches normality is dependent upon the extent of its original departure. If the original distribution had one peak and was not greatly asymmetrical, the distributions of means would approach normality relatively quickly (Fig. 9.7). Under certain conditions, the rate at which a distribution approaches normality can be mathematically determined (Hall 1982). However, the level of math required by these methods prevents their general use.

Weak Law of Large Numbers. If the original data seem so far from

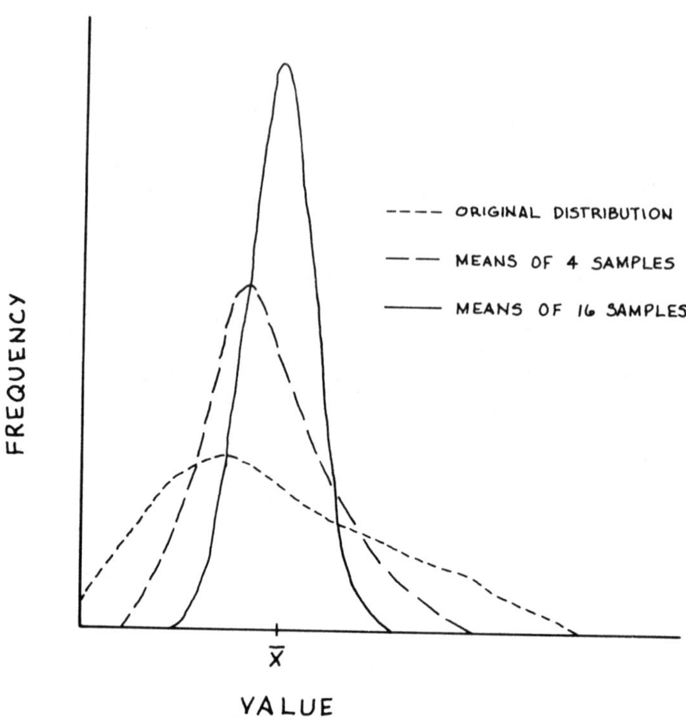

Figure 9.7. Example of the Central Limit Theorem, showing the distributions of means of four and sixteen samples selected at random from an asymmetrical data distribution.

normal that it is unlikely to obtain normality within a reasonable sample size with the Central Limit Theorem, another test for sample adequacy can be used which does not depend on the initial data distribution. This is called the Weak Law of Large Numbers (Mood et al. 1974). It states that the number of samples (N) which must be taken from a population with variance, S^2, such that the mean is within L units of the actual value (for the population) with a probability of 1-Q (Q can be as close to zero as desired, but must be less than one) is:

$$N = \frac{S^2}{L^2 Q} \tag{9}$$

Although this method is more widely applicable than equation (7), it requires as many as three times more samples for the same confidence and accuracy criteria, and therefore should only be considered when the normal distribution can not be achieved.

Estimation of variance. Like the SEM:mean ratio test, both equations (7) and (9) depend upon an accurate estimate of the stand variance. Sowards (1983) presents an excellent discussion of the factors which control variance in vegetation sampling. These include: the uniformity of the mapping unit, the sampling design, and the frequency distribution of the data. He states that the number of samples necessary to estimate the mean is usually less than that necessary to estimate variance. Thus, by the time the variance requirement is met, the test is no longer needed. Testing the accuracy of the variance estimation is possible if the data is very close to normal (Pollard 1977). However, much more work is needed in this area of statistics.

Probabilistic approach. Bonham (1982) proposed a probabilistic approach to sample size which is neither dependent on frequency distribution nor on variance. In this method, the sample size is a function of two parameters, P and W, both between 0 and 1. P is the probability that a certain proportion, W, of the maximum data distribution will fall between the highest and the lowest values of the sample. This sample size is defined as:

$$N = \frac{A \ln[W(1-P)/A] + A - W}{1 - W + A \ln W} \tag{10}$$

where,

$$A = \ln[10/(1-P)] \tag{11}$$

Thus, if one wants to know how many samples would be required to be 80% confident ($P = 0.8$) that 90% ($W = 0.9$) of the possible data points are within the range of values of the sample, this formula would indicate that twenty-nine samples are required. Increasing P to 90% will increase the sample size to thirty-eight.

The sample size indicated by this method is determined entirely from the values of P and W. Bonham points out that these parameters must not be interpreted as being equivalent to the confidence levels of statistical tests. The sample size will always be the same for a particular set of values of P and W, despite the condition of the data. Although Bonham's test is, for this reason, a poor test of sample adequacy, it does add credence to the use of a fixed maximum sample size.

Summary of sample adequacy

The bottom line of sample adequacy is that no single test has been devised which is 100% reliable in its ability to determine the number of samples required for a given degree of confidence. Assumptions and approximations must be made in any test. Figure 9.8 illustrates a flow chart of the methods described in this section, by which a highly reliable estimate of sample size can be calculated. The complexity of this approach precludes its use as a quick field test for adequacy, but most microcomputers could be programmed to perform these analyses.

Perhaps the greatest problem in developing a sample adequacy test for vegetation data is the fact that the actual value for cover and biomass cannot be determined exactly for a stand. Testing a method on actual vegetation requires testing one estimate of these parameters against a different estimate which is considered more accurate. Gaining this added accuracy involves a large investment of manpower. An alternative approach is the use of artificial plant communities for which the actual stand parameters can be determined exactly. This author (Fischer 1982) used computer simulation of line-intercept sampling to compare the predictions of equations (7) and (8) against the actual number of samples required for the level of accuracy prescribed in each equation (L and $P\bar{X}$, respectively). The results (Fig. 9.9) favored the use of a constant value for determining sample adequacy, rather than varying the value according to the coverage of the community. It is hoped that future work in this field will finally resolve the problem of sample adequacy in a manner acceptable to all involved in reclamation.

Determining Success Standards from Pre-mining Vegetation

Methods of controlling for year-to-year variability

Using the methods of vegetation analysis described in the preceding section, it is possible to accurately estimate the pre-mining cover and productivity at a mine site. However, there is a danger in using only a single year's data for establishing success standards. This danger rests in the fact that, like the climate, the cover and productivity of an area varies from year to year. If the success standards are derived from a dry year, bond-release might be granted when revegetation is insufficient.

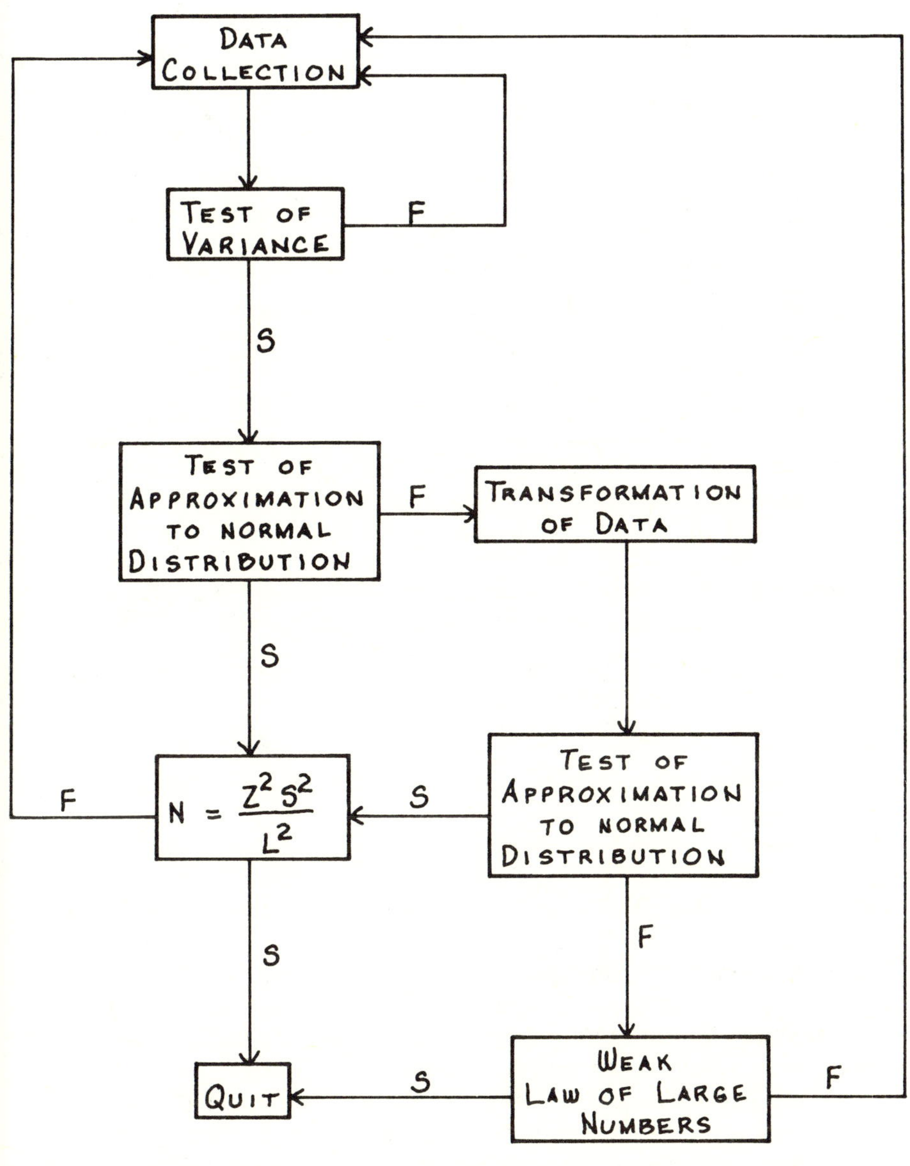

Figure 9.8. Possible flow chart of sample adequacy tests valid for all data types. S and F indicate pathway to be taken when the test is a "success" or "failure," respectively. The test for the approximation to the normal distribution has not been treated here.

　　　　　　　　　N. Timothy Fischer

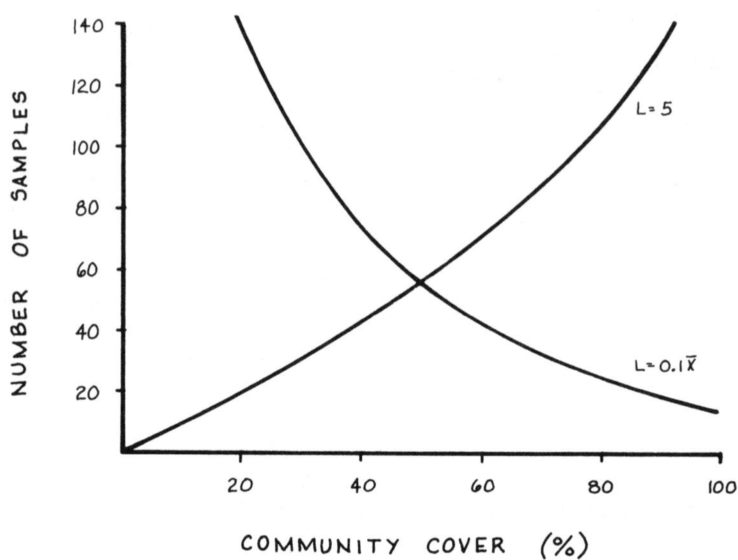

Figure 9.9. A comparison of the number of samples required to achieve sample adequacy according to the formula: $N = \dfrac{Z^2 S^2}{L^2}$ with L held constant and with L as a proportion of the mean. (From Fischer 1982.)

On the other hand, if the sampling year is unusually favorable, the resultant standards might be unattainable in normal years. This section describes some of the methods by which year-to-year variability is taken into account in the development of success standards.

Reference areas.　In the reference-area approach, small stands of vegetation are selected adjacent to the mine site, which represent the vegetation types on the site. These reference areas are selected based upon statistical comparisons with the vegetation types at the mine. Mining can begin immediately after the reference areas are established. At the end of the ten-year liability period, the reference areas are sampled for cover and productivity, and the data are adjusted according to the original extent of the pre-mining vegetation types which they represent. These values are then used as the minimum standards for success.

Several questions have been raised about the use of reference areas. The most significant is whether a small area can truly represent such a large unit as a pre-mining plant community. The larger this original unit is, the more variability it will contain. The small reference area might represent the "average" condition of the unit, but does it represent the extremes, which may respond to climatic variability in quite different ways? A minimum of one reference area per vegetation unit is required by law. If several were selected, there could be greater representation of the variability. However, this becomes cumbersome since each must be procured from the owner of the adjacent land and protected from disturbance through the life of the mine.

Control areas.   Sowards and Hunter (1982) proposed the use of control areas as an alternative to reference areas in order to circumvent the size discrepancy mentioned above. The control area is also a small, off-site stand used as a success standard, but the measurements from it are not directly used as the standard. Instead, a proportion is calculated between the control unit and various units within the pre-mining vegetation, and this proportion is used to make comparisons after revegetation. Thus, one control area can represent several areas within the pre-mining vegetation. In addition, control areas are much more easily located, since exact correspondance to other vegetation is not required.

Historical record.   The historical record is an accepted alternative to the reference area in New Mexico. In this approach, the pre-mining vegetation on the site is sampled for several consecutive years (seven in New Mexico). From these data, an idea of the year-to-year variation in cover and productivity is obtained. This method has the advantage of basing the standards completely on the vegetation of the site; however, it requires that the whole site be sampled each year of the prescribed period, making it more labor intensive than other methods.

The use of models to reduce sampling

This author believes that the principles in this and previous chapters can be applied to reach an ecologically sound and cost-effective middle ground among the methods described above. This approach would begin with an intensive vegetation survey within the context of ecological response units (ERUs), as identified and described according to Bonham et al. (1980). The entire site is first subdivided into a grid of equal-size, square units, each of which is classified according to range site (soil type), slope, and orientation. Adjacent units of the same type are combined into larger ERU blocks.

Ideally, each ERU block would be sampled during the first year of data collecton; however, when many blocks of the same type are present, randomly selected blocks could be sampled. Sampling is confined by the boundaries of individual blocks and is done until adequacy has been reached for each block. It is suggested that this be done with relatively small sampling units (for example, short lines or small quadrats). This will provide good coverage within the block, giving better representation of its internal variability. Although small sampling units cause a larger sample variance, thus requiring more samples to achieve adequacy, each can be performed more quickly. Furthermore, the uniformity of the ERU will keep the variance low.

Admittedly, the initial sampling will entail a large investment of manpower, but no more than the amount of effort used for pre-mining assessments of reclamation potential (see Chapter 3). Data of this nature will also enable the application of predictive models (Chapter 5) to replace actual field sampling as more information is learned about the system. For example, multiple regression can be used to determine the relationships between the habitat characteristics of the ERU (soil, slope, and orientation) and the vegetation parameters (cover, productivity, and so forth). From this analysis, certain blocks might be identified which can be used as on-site control areas, from which measurements can be used to program the model to predict the parameter values for other blocks. Thus, after an initial bout of intense sampling, all effort could be confined to the control blocks for the purpose of establishing a historical record.

Other considerations might reduce the amount of sampling even more. Some ERUs, for example, might be eliminated from future sampling on the grounds that their habitat will not be part of the array of habitats on the post-mining surface. Badlands, for example, are highly erosive (Chapter 2) and have very low plant coverages. Such an area on the reclaimed surface would be unacceptable, therefore, sampling of badland habitats after the first year would be unnecessary.

Methods for reducing productivity sampling

Still another way to reduce the amount of sampling is to establish relationships between vegetation parameters that will enable predictions instead of fieldwork in the later years of a historical record. Productivity, for example, is often the most time-consuming parameter to measure; also, intensive sampling for this parameter should be avoided because vegetation must be clipped, and hence destroyed. *Double sampling* is a method currently used to reduce the amount of clipping required for these measurements (Bonham et al. 1980). In this method, the productivity within the quadrat is estimated by the investigator. Periodically, the estimate is checked by clipping and weighing the plants in the quadrat. By determining the relationship (by linear regression) between the estimates and the clippings, the accuracy of the other estimates can be determined and adjustments made. The methods proposed herein are simply extensions of the idea of double sampling, which is accepted by the regulatory agencies for the characterization of productivity.

Hargis et al. (1983) proposed that productivity can be predicted from cover data through the use of regression lines. This results in a great saving of time in the collection of vegetation data. By using this approach, productivity would only have to measured periodically through the sampling period, or perhaps only in the last year when vegetation destruction will not be a consideration.

Summary: Do the current methods satisfy the objectives of reclamation?

The current tests for revegetation success emphasize the effectiveness of the vegetation in controlling erosion and in supporting the prescribed post-mining land use. This is done by comparing cover and productivity measurements from the reclaimed surface with the standards for these parameters derived from reference areas, historical records, or other accepted methods. The tests for success are performed once at the end of the liability period.

The question must be asked, however, whether this legal definition of success actually supports the ultimate goals of reclamation, that is, the protection of the quality of the environment and the utility of the land. Only time will tell whether the reclamation efforts on the large surface mines in the San Juan Basin have achieved these goals. Will the vegetation remain stable on "successfully" reclaimed surfaces over the decades beyond bond-release? How will the implementation of the post-mining land use affect the vegetation? How will plant succession—that is, the natural progression of a plant community toward the highest state of compatibility with its environment—affect the character of the vegetation in the future? These are questions for which hard evidence does not yet exist to provide answers.

This chapter has described some of the basic methods by which cover and production of the pre-mining vegetation can be measured accurately and efficiently in order to establish standards for revegetation success, but it has not considered the ecological validity of these measurements.

Vegetation is not merely an assemblage of plants, but a complex, interactive system intregrated into the environment. The stability of this system is determined by the interactions among its internal elements, including the vegetation, the soil, and the soil biota. The determination of revegetation success must take this fact into account, but the current methods do not. The attainment of a certain cover and productivity, no matter what the values are, does not indicate stability. The ten-year liability period which is required before success can be tested partially fulfills this need; still, it is only the *condition* of the vegetation at the end of this period which is measured, and not the trend which describes the natural succession of the plant commmunity. Ideally, the evaluation of revegetation success must look beyond the vegetation to assess the environment which has been created to support the vegetation. Evaluations must also anticipate possible changes or potentials for changes in the condition of the reclaimed environment over time.

Presently, we are far from this level of sophistication in the analysis of vegetation on mine sites, but the recent and rapid proliferation of ideas on this facet of reclamation science suggests that the gap may be closed sooner than expected.

References

Anderson, E. W. 1983. Ecological site/ range site/ habitat site — a viewpoint. Rangelands 5(4): 187-188.

Bonham, C. D. 1982. Approach to sample adequacy. *In* E. F. Aldon and W. R. Oaks, eds., Reclamation of mined lands in the Southwest, 8-10. Albuquerque, NM: New Mexico Chapter, Soil Conservation Society of America.

Bonham, C. D., L. L. Larson, and A. Morrison. 1980. A survey of techniques for the measurement of herbaceous and shrub production, cover and diversity on coal lands in the West. Office of Surface Mining, Region 5, Denver, CO.

Brady, N. C. 1974. The nature and properties of soils. 8th edition. New York, NY: MacMillan Publishing Co.

Brown, D. E., C. H. Lowe, and C. P. Pase. 1979. A digitized classification system for the biotic communities of North America, with community (series) and association examples for the Southwest. Journal of the Arizona-Nevada Academy of Science 14(supp. 1): 1-16.

Daubenmire, R. F. 1968. Plant communities: a textbook of plant synecology. New York, NY: Harper and Row.

Fischer, N. T. 1982. An evaluation of sample adequacy for pre- and postmine vegetation surveys using computer-generated artificial communities. *In* 1982 Symposium on surface mining hydrology, sedimentology, and reclamation, 103-110. Lexington, KY: University of Kentucky, Lexington, KY.

Gleason, H. A. 1939. The individualistic concept of the plant association. American Midland Naturalist 21:92-110.

Hall, P. 1982. Rates of convergence in the central limit theorem. Boston, MA: Pitman Advanced Publishing Program.

Hargis, N. E., E. F. Redente, W. E. Sowards, and D. G. Steward. 1983. Eliminating biomass sampling as a requirement for evaluating revegetation success by predicting biomass from cover. *In* E. F. Redente, W. E. Sowards, D. G. Steward, and T. L. Ruiter, eds., Symposium on western coal mining regulatory issues: land use, revegetation, and management, 93-97. Range Science Department science series no. 35. Fort Collins, CO: Colorado State University.

Keammerer, W. R. 1983. Measures of sample adequacy. *In* E. F. Redente, W. E. Sowards, D. G. Steward, and T. L. Ruiter, eds., Symposium on western coal mining regulatory issues: land use, revegetation, and management, 72-74. Range Science Department science series no. 35. Fort Collins, CO: Colorado State University.

Komarkova, V. 1983. Comparison of habitat type classification to some other classification methods. *In* W. H. Moir and L. Hendzel, eds., Proceedings of the workshop on southwestern habitat types, 21-31. Albuquerque, NM: Southwest Region, U.S. Forest Service.

McIntosh, R. P. 1967. The continuum concept of vegetation. Botanical Review 33: 130-187.

Mood, A. M., F. A. Graybill, and D. C. Boes. 1974. Introduction to the theory of statistics. 3rd edition. New York, NY: McGraw-Hill.

Mueller-Dombois, D., and H. Ellenberg. 1974. Aims and methods of vegetation ecology. New York, NY: John Wiley and Sons.

Oosting, H. J. 1956. The study of plant communities. 2nd edition. San Francisco, CA: W. H. Freeman and Co.

Pielou, E. C. 1977. Mathematical ecology. New York, NY: John Wiley and Sons.

Pollard, J. H. 1977. A handbook of numerical and statistical techniques. Cambridge, England: Cambridge University Press.

Snedecor, G. W., and W. G. Cochran. 1967. Statistical methods. Ames, IA: Iowa State University Press.

Sowards, W. E. 1983. Achieving an adequate sample size for vegetation analysis. *In* E. F. Redente, W. E. Sowards, D. G. Steward, and T. L. Ruiter, eds., Symposium on western coal mining regulatory issues: land use, revegetation, and management, 63-71. Range Science Department science series no. 35. Fort Collins, CO: Colorado State University.

Sowards, W. E., and D. H. Hunter. 1982. A critical analysis of revegetation standards for coal mined lands. *In* E. F. Aldon and W. R. Oaks, eds., Reclamation of mined lands in the Southwest, 201-212. Albuquerque, NM: New Mexico Chapter, Soil Conservation Society of America.

Warming, E. 1909. Oecology of plants. Translated by P. Groom and I. B. Balfour. Oxford: Clarendon Press.

Whittaker, R. H. 1956. Vegetation of the Great Smoky Mountains. Ecological Monographs. 26: 1-80.

Whittaker, R. H. 1962. Classification of natural communities. Botanical Review 28: 1-239.

Whittaker, R. H., ed., 1978. Ordination of plant communities. The Hague, The Netherlands: Dr. W. Junk.

Notes on Contributors

Earl F. Aldon is Project Leader for the Albuquerque Unit of the Rocky Mountain Forest and Range Experiment Station, which is a research arm of the U.S. Forest Service (Department of Agriculture). Mr. Aldon received an MS in Hydrology from the University of Michigan in 1953. His studies of watershed management and range rehabilitation have been widely published in such journals as *Reclamation and Revegetation*, the *Journal of Range Management*, and the *Journal of Environmental Quality*.

Ned Z. Elkins is Manager of Environmental Affairs for Sunbelt Mining Company, a subsidiary of Public Service of New Mexico. Dr. Elkins received Doctorates in Environmental Biology and Civil Engineering from New Mexico State University in 1982. He has developed reclamation plans and obtained operations permits for a number of mines in the West, including the first gold-leaching facility in the Sierra Nevada (California). Dr. Elkins has publications in *Oecologia*, the *Journal of Arid Environments*, and several other sources.

N. Timothy Fischer is an Environmental Scientist for IT Corporation at the Waste Isolation Pilot Plant (WIPP), which is an R&D nuclear waste facility in southeastern New Mexico. Mr. Fischer received an MS in Plant Ecology from the University of New Mexico in 1984. His responsibilities at the WIPP project include ecological research, radiological surveillance, and land rehabilitation. His previous research efforts have encompassed archaeology, wetland ecology, and methods of environmental inventory.

William W. Fuller is the State Agronomist and Plant Materials Specialist with the Soil Conservation Service in New Mexico. Dr. Fuller received a PhD in Crop Science from the University of Oklahoma in 1972. He is currently directing the program for Arid Range Seeding and Equipment Techniques (ARSET), wherein methods and materials are being imported from western Australia in order to improve select North American vegetation types.

Loren D. Potter is Professor Emeritus of Plant Ecology, Biology Department, University of New Mexico. Dr. Potter received his PhD at the University of Minnesota in 1948. He served as Department Chairman and Professor of Plant Ecology at UNM for 27 years, doing research in a variety of ecological problems applied to environmental science. He has published in many ecological journals and is coauthor of *Water Resources in the Southern Rockies and High Plains: Forest Recreational Use and Aquatic Life* and *New Mexico Grasses: A Vegetative Key.*

Charles C. Reith is an Environmental Scientist with International Technology Corporation. Dr. Reith received his PhD in Ecology at the University of New Mexico in 1984. His research on reclamation includes an analysis of reclamation planning in the federal coal leasing program, performed for the Office of Technology Assessment (US Congress). He currently directs environmental field activities for the Waste Isolation Pilot Plant, a nuclear waste facility in southeastern New Mexico.

David G. Scholl is a Scientist with the Rocky Mountain Forest and Range Experiment Station of the US Forest Service (Department of Agriculture). Mr. Scholl received an MS in Soil Science from Michigan State University in 1965. His research emphasizes water relations in mine spoils and heavily grazed rangelands in order to develop long-term rehabilitation plans. His publications have appeared in *Reclamation and Revegetation, Soil Science Society of America,* and other journals.

Stephen G. Wells is an Associate Professor of Geology at the University of New Mexico. Dr. Wells received a PhD in Geology at the University of Cincinnati in 1976. For the last seven years, he has conducted a supervised research on geomorphological aspects of reclamation and waste-siting in the arid Southwest. Dr. Wells recently organized the 1983 conference of the American Geomorphological Field Group.

Walter G. Whitford is a Professor of Biology at New Mexico State University. Dr. Whitford received his PhD in Zoology from the University of Rhode Island in 1964. He has been Principal Investigator on several long-term ecological research programs of major significance at the Jornada Experimental Range near Las Cruces, NM, and he has published more than a hundred papers in journals such as *Ecological Monographs* and *Oecologia.* Dr. Whitford recently received the Westhaver Award for outstanding research at NMSU.

Index